*Konrad Zilch*
*Roland Niedermeier*
*Wolfgang Finckh*

**Strengthening of Concrete Structures with Adhesively Bonded Reinforcement**

*Konrad Zilch*
*Roland Niedermeier*
*Wolfgang Finckh*

# Strengthening of Concrete Structures with Adhesively Bonded Reinforcement

Design and Dimensioning

of CFRP Laminates and Steel Plates

**The Authors**

***Prof. Dr.-Ing. habil. Dr.-Ing. E. h. Konrad Zilch***
Technische Universität München
Theresienstr. 90
80333 Munich
Germany

***PD Dr.-Ing. habil. Roland Niedermeier***
Technische Universität München
MPA BAU
Theresienstr. 90
80333 Munich
Germany

***Dr.-Ing. Wolfgang Finckh***
Wayss & Freytag Ingenieurbau AG
Eschborner Landstraße 130-132
60489 Frankfurt/Main
Germany

**The Editors**

***Prof. Dipl.-Ing. Dr.-Ing. Konrad Bergmeister***
Ingenieurbüro Bergmeister
Peter-Jordan-Straße 113
1190 Vienna
Austria

***Dr.-Ing. Frank Fingerloos***
German Society for Concrete and Construction Technology
Kurfürstenstr. 129
10785 Berlin
Germany

***Prof. Dr.-Ing. Dr. h. c. mult.***
Johann-Dietrich Wörner
German Aerospace Center
Linder Höhe
51145 Cologne
Germany

**Coverphoto:** Ludwig Freytag GmbH & Co. KG, Oldenburg, Germany

**Coverdesign:** Hans Baltzer, Berlin, Germany

The original German text is published in Beton-Kalender 2013, ISBN 978-3-433-03000-4 and titled "*Geklebte Verstärkung mit CFK-Lamellen und Stahllaschen*".



**Library of Congress Card No.:** applied for

**British Library Cataloguing-in-Publication Data**
A catalogue record for this book is available from the British Library.

**Bibliographic information published by the Deutsche Nationalbibliothek**
The Deutsche Nationalbibliothek lists this publication in the Deutsche Nationalbibliografie; detailed bibliographic data are available on the Internet at <http://dnb.d-nb.de>.



**Print ISBN:** 978-3-433-03086-8
**ePDF ISBN:** 978-3-433-60403-8
**ePub ISBN:** 978-3-433-60405-2
**mobi ISBN:** 978-3-433-60402-1
**oBook ISBN:** 978-3-433-60401-4

**Typesetting:** Thomson Digital, Noida, India

**Printing and Binding:** betz-druck GmbH, Darmstadt, Germany

Printed in the Federal Republic of Germany
Printed on acid-free paper

# Contents

# Editorial

The *Concrete Yearbook* is a very important source of information for engineers involved in the planning, design, analysis and construction of concrete structures. It is published on a yearly basis and offers chapters devoted to various, highly topical subjects. Every chapter provides extensive, up-to-date information written by renowned experts in the areas concerned. The subjects change every year and may return in later years for an updated treatment. This publication strategy guarantees that not only is the latest knowledge presented, but that the choice of topics itself meets readers' demands for up-to-date news.

For decades, the themes chosen have been treated in such a way that, on the one hand, the reader gets background information and, on the other, becomes familiar with the practical experience, methods and rules needed to put this knowledge into practice. For practising engineers, this is an optimum combination. In order to find adequate solutions for the wide scope of everyday or special problems, engineering practice requires knowledge of the rules and recommendations as well as an understanding of the theories or assumptions behind them.

During the history of the *Concrete Yearbook*, an interesting development has taken place. In the early editions, themes of interest were chosen on an ad hoc basis. Meanwhile, however, the building industry has gone through a remarkable evolution. Whereas in the past attention focused predominantly on matters concerning structural safety and serviceability, nowadays there is an increasing awareness of our responsibility with regard to society in a broader sense. This is reflected, for example, in the wish to avoid problems related to the limited durability of structures. Expensive repairs to structures have been, and unfortunately still are, necessary because in the past our awareness of the deterioration processes affecting concrete and reinforcing steel was inadequate. Therefore, structural design should now focus on building structures with sufficient reliability and serviceability for a specified period of time, without substantial maintenance costs. Moreover, we are confronted by a legacy of older structures that must be assessed with regard to their suitability to carry safely the increased loads often applied to them today. In this respect, several aspects of structural engineering have to be considered in an interrelated way, such as risk, functionality, serviceability, deterioration processes, strengthening techniques, monitoring, dismantlement, adaptability and recycling of structures and structural materials plus the introduction of modern high-performance materials. The significance of sustainability has also been recognized. This must be added to the awareness that design should focus not just on individual structures and their service lives, but on their function in a wider context as well, i.e. harmony with their environment, acceptance by society, responsible use of resources, low energy consumption and economy. Construction processes must also become cleaner, cause less environmental impact and pollution.

The editors of the *Concrete Yearbook* have clearly recognized these and other trends and now offer a selection of coherent subjects that reside under the common "umbrella" of a broader societal development of great relevance. In order to be able to cope with the corresponding challenges, the reader can find information on progress in technology, theoretical methods, new research findings, new ideas on design and construction, developments in production and assessment and conservation strategies. The current selection of topics and the way they are treated makes the *Concrete Yearbook* a splendid opportunity for engineers to find out about and stay abreast of developments in engineering knowledge, practical experience and concepts in the field of the design of concrete structures on an international level.

Prof. Dr. Ir. Dr.-Ing. h. c. *Joost Walraven*, TU Delft
Honorary president of the international concrete federation *fib*

# 1 Introduction

## 1.1 The reason behind this book

The main reason is the revised approach to the design of adhesively bonded strengthening measures for concrete members given in the guideline [1] (q.v. [2]) published by the Deutscher Ausschuss für Stahlbeton DAfStb (German Committee for Structural Concrete). This book explains the design rules of the DAfStb guideline, together with their background, and uses examples to illustrate their use. The scope of the explanations and background information provided here is mainly based on works that have already been published. However, some rules that so far have been dealt with in detail in committee meetings only are elaborated here for the first time.

## 1.2 Strengthening with adhesively bonded reinforcement

The strengthening of concrete members means using constructional measures to restore or improve their load-carrying capacity, serviceability, durability or fatigue strength. The effects of strengthening measures can generally be described in quantitative terms and therefore analysed numerically. Besides numerous other methods (see [3, 4], for example), the subsequent strengthening of existing concrete members can be achieved by using adhesives to bond additional reinforcing elements onto or into those members. This topic of reinforcement bonded with adhesive has been the subject of many contributions to various editions of the *Beton-Kalender* in the past (see [5, 6]). However, design approaches for adhesively bonded reinforcement have continued to evolve (see [7, 8]) and the new DAfStb guideline [1, 2] on this subject revises those design methods and adapts them to our current state of knowledge. In principle, the DAfStb guideline together with a corresponding system approval allows the following concrete member strengthening measures to be carried out:

- Flexural strengthening with externally bonded (surface-mounted) CFRP strips, CF sheets and steel plates
- Flexural strengthening with CFRP strips bonded in slots (near-surface-mounted reinforcement)
- Shear strengthening with externally bonded CF sheets and steel plates
- Column strengthening with CF sheets as confining reinforcement.

Figure 1.1 provides an overview of these methods. The term 'adhesively bonded' is used in this book as universal expression comprising both methods 'externally bonded' and 'near-surface-mounted'.

*Strengthening of Concrete Structures with Adhesively Bonded Reinforcement: Design and Dimensioning of CFRP Laminates and Steel Plates*. First Edition. Konrad Zilch, Roland Niedermeier, and Wolfgang Finckh.

**Fig. 1.1** (a) Externally bonded and near-surface-mounted CFRP strips; (b) flexural strengthening with externally bonded CFRP strips together with shear strengthening in the form of externally bonded steel plates (photo: Laumer Bautechnik GmbH); (c) column strengthening with CF sheets as confining reinforcement (photo: Laumer Bautechnik GmbH)

# 2 DAfStb guideline

## 2.1 The reasons for drawing up a guideline

In the past, the product systems as well as the design and installation of adhesively bonded reinforcement were regulated in Germany by national technical approvals and individual approvals. Such approvals contained provisions covering the materials, the design of the strengthening measures, the work on site and the monitoring of products. There were several reasons why it was deemed necessary to revise the design approaches of the earlier approvals.

One of those reasons was the harmonization of standards across Europe, leading to national standards and regulations being successively adapted to the European standards. These developments also render it necessary to adapt the former national approvals to the new generation of standards.

Furthermore, the results of numerous research projects carried out in recent years had only been partly incorporated in the older regulations, which therefore no longer matched the current state of knowledge. Therefore, industry, the building authorities and the German Research Foundation (DFG) made substantial funds available for researching adhesively bonded reinforcement. That led to many scientific projects in the German-speaking countries and adhesively bonded reinforcement gradually becoming a standard method in the building industry. Consequently, all the groups involved regarded the preparation of a universal guideline as indispensable.

## 2.2 Preparatory work

In order to produce a universal guideline reflecting the current level of knowledge, the German Committee for Structural Concrete (DAfStb) first commissioned a report on the current situation [7] to document and collate national and international knowledge. A database of test results containing almost all the experimental studies carried out nationally and internationally was also set up and compared with the established models and the guidelines available elsewhere in the world. During the drafting of the report it became apparent that the knowledge necessary to produce an effective guideline was lacking in some areas. Therefore, under the direction of the DAfStb, a research project was initiated in which all the groups interested took part. The research work was carried out by the technical universities in Munich and Brunswick, both of which had been working continually on adhesively bonded reinforcement for more than 20 years. The project was financed by the owners of the approvals (Bilfinger Berger AG, Laumer Bautechnik GmbH, Ludwig Freytag GmbH & Co. KG, MC-Bauchemie Müller GmbH & Co. KG, S&P Clever Reinforcement Company AG, Sika Deutschland GmbH, Stocretec GmbH), the Federal Institute for Research on Building, Urban Affairs & Spatial Development (BBSR) plus a number of associations and consulting engineers. Issues surrounding the bond strength under static loads [9] and dynamic loads [10] plus the shear strength [11] were successfully clarified during this project.

*Strengthening of Concrete Structures with Adhesively Bonded Reinforcement: Design and Dimensioning of CFRP Laminates and Steel Plates*. First Edition. Konrad Zilch, Roland Niedermeier, and Wolfgang Finckh.

## 2.3 Work on the guideline

A subcommittee set up by the DAfStb began drafting the guideline as the research work progressed. In accordance with DIN 820-1 [12], the groups involved (building authorities, industry, research centres, official bodies, trade associations) were all represented equally on the subcommittee. Within a year, a draft version had been prepared. The draft, incorporating the results of research projects but also the experiences of the members of the subcommittee, appeared in March 2011 as a paper for discussion and was announced and presented to the industry in numerous publications [13–19]. Comments and objections could be filed by mid-September 2011. A meeting to discuss and decide on objections was subsequently held. Following its notification by the European Union, the finished guideline became available in the summer of 2012. It can be purchased from Beuth Verlag. The DAfStb guideline [2] is also available in English.

## 2.4 The structure and content of the guideline

### 2.4.1 General

The DAfStb guideline covering the strengthening of concrete members with adhesively bonded reinforcement [1, 2] provides rules for design and detailing, the application of national technical approvals for strengthening systems, execution and additional rules for planning strengthening measures employing adhesively bonded reinforcement.

The guideline is divided into four parts. The first part covers the design and detailing of strengthening measures using reinforcement bonded with adhesive. This part of the guideline supplements DIN EN 1992-1-1 [20] with its associated National Annex [21] by providing the additional requirements necessary for adhesively bonded reinforcement. The second part of the guideline, together with the system approvals, describes the products and systems used for strengthening measures with adhesively bonded reinforcement. The third part covers the execution, and also contains advice on installing the specified strengthening measures. The fourth part of the guideline contains additional rules for planning strengthening measures.

### 2.4.2 Design and detailing

As mentioned above, the first part of the DAfStb guideline supplements DIN EN 1992-1-1 [20] and its associated National Annex [21]. Its structure corresponds exactly to that of DIN EN 1992-1-1, and there are additional provisions for materials, durability, ultimate limit state, serviceability limit state, reinforcing principles and detailing.

Chapters 3, 5 and 7 of this book describe the design and detailing provisions for different strengthening measures and the background to these.

### 2.4.3 Products and systems

The second part of the DAfStb guideline covers the application of system approvals for strengthening measures employing adhesively bonded reinforcement. Strengthening must be carried out with an approved strengthening system using strengthening products to DIN EN 1504-1 [22]. A strengthening system consists of various properly matched construction products whose usability as components in a strengthening system must be verified within the scope of a national technical approval for the strengthening system.

The main elements of such a strengthening system are:

- the strengthening elements made from carbon fibre materials (CFRP strips or CF sheets) or steel flats/angles,
- the adhesive,
- a primer based on epoxy resin as a component for protecting steel parts against corrosion, and
- a repair mortar based on epoxy resin which includes a bonding agent.

### 2.4.4 Execution

The third part of the DAfStb guideline deals with the work on site. It contains advice and provisions for carrying out the strengthening measures. For example, it provides information on the pretreatment of members and the associated inspections to be carried out. In addition, it specifies the requirements to be met by contractors who carry out strengthening measures.

### 2.4.5 Planning

The fourth part of the DAfStb guideline contains supplementary regulations for planning strengthening measures. It includes definitions of the requirements to be satisfied by the member being strengthened. There are also recommendations regarding the scope of the planning and the measures required to determine the actual condition of the member to be strengthened. In addition, all design and construction work must take into account the DAfStb's guideline on the maintenance of reinforced concrete [23].

## 2.5 Safety concept

As with DIN EN 1992-1-1 [20] and its associated National Annex [21], the DAfStb guideline is based on the safety concept of DIN EN 1990 [24] together with its National Annex [25].

The guideline specifies partial safety factors for externally bonded reinforcement, which are given in Table 2.1. A distinction is made between the partial safety factors for the strength of the bonded reinforcement and those for the bond of the bonded reinforcement.

The partial safety factors for the strength of externally bonded reinforcement were chosen according to the *fib* recommendations [26]. The partial safety factor proposed in [26] for CFRP strips has also been evaluated statistically by Blaschko [27] and used in the design rules of earlier approvals [28, 29].

The partial safety factors for the bond of reinforcement attached with adhesive depend on the mode of failure. In the case of near-surface-mounted reinforcement and when bonding steel to steel or CFRP to CFRP, it is generally the adhesive that governs a bond failure, and the safety factor for the adhesive according to [27] is used, as it was already the case in the earlier approvals (see [29], for example).

**Table 2.1** Partial safety factors for adhesively bonded reinforcement for the ultimate limit state.

| Design situation | CFRP strips | CF sheets | Bond of externally bonded reinforcement | Bond of near-surface-mounted reinforcement | Bond of steel on steel or CFRP on CFRP |
|---|---|---|---|---|---|
| Designation | $\gamma_{LL}$ | $\gamma_{LG}$ | $\gamma_{BA}$ | $\gamma_{BE}$ | $\gamma_{BG}$ |
| Persistent and transient | 1.2 | 1.35 | 1.5 | 1.3 | 1.3 |
| Accidental | 1.05 | 1.1 | 1.2 | 1.05 | 1.05 |

A bond failure with externally bonded reinforcement (steel plates, CFRP strips, CF sheets) entails a failure in the layer of concrete near the surface. For this reason, when it comes to the bond of externally bonded reinforcement, the DAfStb guideline uses the partial safety factors for concrete failure according to [20], as in the earlier approvals [28]. According to [30], using such a global partial safety factor for the bond of externally bonded reinforcement – instead of individual partial safety factors for every variable in the design equations – resulted in the smallest deviations in the level of safety when comparing various sample calculations.

## 2.6 Applications

### 2.6.1 Member to be strengthened

The DAfStb guideline can be applied to concrete members complying with DIN EN 1992-1-1 [20, 21]. The design approaches of the guideline were prepared based on the theories of mechanics and calibrated and validated by means of tests on normal-weight concretes of strength classes C12/15 to C50/60. Therefore, the design approaches of the guideline should not be applied to other construction materials without carrying out additional investigations. In order to keep within the range of experience of the experimental studies available, the guideline specifies the following exceptions:

- The DAfStb guideline cannot be used for strengthening lightweight concrete.
- The DAfStb guideline only covers the strengthening of normal-weight concretes of strength classes C12/15 to C50/60.

Furthermore, the guideline [1, 2] cannot be applied to components made from steel fibre-reinforced concrete or autoclaved aerated concrete. So extending DIN EN 1992-1-1 [20] by means of the DAfStb guideline on steel fibre-reinforced concrete [31] and combining this with the DAfStb guideline on the strengthening of concrete members with externally bonded reinforcement [1, 2] is not permitted.

The DAfStb guideline on the strengthening of concrete members with adhesively bonded reinforcement [1, 2] can only be used to design strengthening for concrete components. It cannot be used to design strengthening measures for masonry components (see [32], for example), timber components (see [33], for example), steel components (see [34], for example) or composite components, nor does it cover combinations with other methods of strengthening.

### 2.6.2 Strengthening systems

The DAfStb guideline [1, 2] covers strengthening systems incorporating adhesively bonded reinforcement. German construction law requires that the strengthening system being used must have been granted a national technical approval.

Externally bonded reinforcement assumes that an adhesive, based on an epoxy resin, is used to attach reinforcing elements in the form of steel or carbon fibre (CF) materials to a concrete substrate from which all intrinsic substances (cement laitance) and foreign matter (plaster, render, paint) have been removed using suitable methods. An externally bonded strengthening solution therefore assumes that a compact tension member, linear elastic in the area of the stress–strain curve considered, is attached to the concrete with the help of a high-strength adhesive. Owing to this high-strength adhesive and the compact form of the reinforcing element, a concrete failure is always assumed in the case of the debonding of externally bonded reinforcement. For these reasons, the design approaches cannot be directly transferred to other forms of strengthening, e.g. upgrading with textile-reinforced concrete (see [35], for example). Most of the experiments carried out and practical experience gained so far in Germany has involved strengthening with steel plates and CFRP strips. Considerable experience has been gained with externally bonded CF sheets, too. However, there can be much greater differences between different CF sheet products than is the case with CFRP strips or steel plates. Therefore, the DAfStb guideline only specifies bond values for CFRP strips or steel plates. Owing to the mechanics background to the design equations, it is readily possible, however, to transfer the design approaches to CF sheets as well by adapting or verifying the bond values given in the guideline. These bond values are included in the national technical approvals for CF sheets.

When it comes to near-surface-mounted reinforcement (i.e. reinforcement bonded in slots), it is assumed that reinforcing elements in the form of prefabricated CFRP strips are bonded with an epoxy resin adhesive in slots sawn or milled in the concrete cover. It is not possible to use steel elements instead of CFRP slots in such cases because assuring adequate corrosion protection is awkward. Likewise, it is not possible to apply the design approaches to other types of reinforcement, e.g. round bars, because of the dissimilar bond behaviour.

Besides conventional bonding, it is also possible to use appropriate equipment to attach prestressed CFRP strips (see [36–38], for example). However, owing to the numerous unanswered questions about prestressed bonded reinforcement, the DAfStb guideline [1, 2] does not include any design approaches for this form of strengthening.

### 2.6.3 Ambient conditions

As the acceptable environmental conditions depend heavily on the properties of the strengthening system, the DAfStb guideline [1, 2] provides only general advice on this aspect. The permissible ambient conditions, such as exposure classes and other environmental influences, are – like the ensuing protective measures – regulated in the national technical approvals for the strengthening systems.

Generally, without additional protective measures, externally bonded reinforcement may only be used for exposure classes X0, XC1 or XC3 to DIN EN 1992-1-1 Table 4.1. In addition, the members in the region of the bonded reinforcement may not be exposed to strong UV radiation (direct sunshine or indirect sunshine reflected off snow or water) or alternating or permanent saturation.

One special aspect of bonded strengthening systems is their sensitivity to elevated temperatures. Cold-curing epoxy resin adhesives are normally used for retrofitted strengthening measures. These thermosetting polymers are amorphous and very stable below a certain temperature. However, at higher temperatures the crystalline phase gradually breaks up and the adhesive loses its strength over the glass transition range. The guideline therefore specifies that no loads may be allocated to the externally bonded reinforcement once the start of the glass-liquid transition (minus a safety margin) has been reached. This temperature is denoted $T_f$ and owing to its dependence on the particular product is noted in the associated national technical approval. Without heat treatment, this figure lies between 40 and 60 °C for the epoxy resin adhesives currently on the market. According to the current state of knowledge, it is also known that the glass transition temperature essentially depends on the temperature during bonding and curing and during any further hardening involving intensive heating. It was observed in [39], for example, that a static glass transition temperature is not acceptable as a thermal serviceability limit because it is heavily dependent on the curing conditions. Therefore, care should be taken to ensure that when bonding at the bottom end of the service temperature range, no abrupt rise in temperature to the top end of the glass transition range can take place.

### 2.6.4 Fire protection

As described in the preceding section, adhesively bonded reinforcement is especially sensitive to elevated temperatures. Fire protection should therefore be paid special attention. Basically, the options are either carrying out a structural fire analysis without taking into account the bonded reinforcement or protecting the bonded reinforcement against heat by applying a suitable protective system. Examples of structural fire analyses for members with bonded reinforcement can be found in [40], for example.

A structural fire analysis, ignoring the adhesively bonded reinforcement, can be carried out according to DIN EN 1992-1-2 [41] in conjunction with its National Annex [42]. However, this is generally associated with the degree of strengthening being limited to some extent. In addition, through using an approved fire protection system to protect the internal reinforcement it is often possible to verify that the member has an adequate load-carrying capacity in the event of a fire even in the case of failure of the bonded reinforcement.

The other possibility is to protect the adhesively bonded reinforcement with a fire protection system that is approved for that particular reinforcement. At the time of drawing up the DAfStb guideline, however, no systems for protecting bonded reinforcement had been granted a national technical approval. Fire tests (see [43]) on members with near-surface-mounted reinforcement and fire protection systems have revealed that the protection of bonded reinforcement must satisfy much higher requirements than is the case with conventional systems.

## 2.7 Relationship with other regulations

The DAfStb guideline for the strengthening of concrete members with adhesively bonded reinforcement [1, 2] must be seen in the context of all national and European standards that cover product, design and construction aspects. In principle, the DAfStb guideline supplements DIN EN 1992-1-1 [20] and its associated National Annex [21]. Strengthening requires an approved system with products to DIN EN 1504-1 [22]. In addition, design and construction work must take account of the DAfStb guideline on the protection and maintenance of concrete members [23]. Every contractor intending to perform strengthening measures must be properly qualified, which in Germany requires proof of suitability. Figure 2.1 provides a basic overview of the DAfStb guideline in the context of the other European and German standards/ guidelines plus system approvals.

However, as many strengthening projects involve members that cannot be assessed using the latest design codes, and concretes or reinforcing steels that do not comply with current standards, it may be necessary to deviate from the framework of standards on occasions. When as-built documents are available, the material parameters may be taken from old standards, for instance, and converted to current reference figures by applying suitable methods (see [45–48], for example).

In individual cases it will also be necessary to investigate whether the presence of certain construction products, e.g. prestressing steel sensitive to stress cracking corrosion (see [49–52], for example), represents a risk to the overall safety of the member.

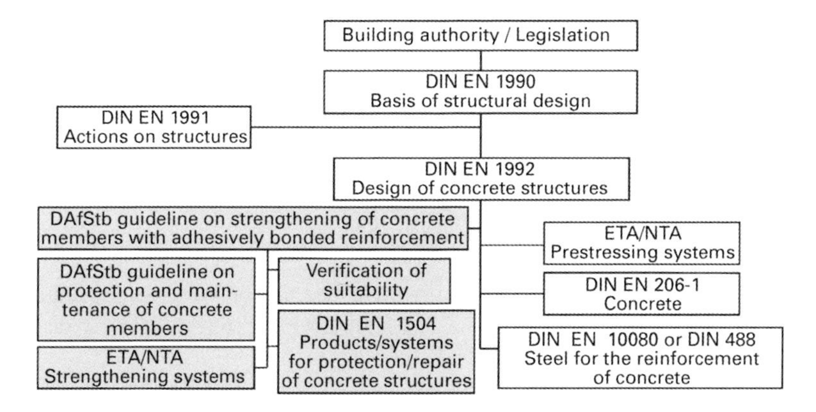


**Fig. 2.1** The DAfStb guideline in the context of German and European design, product and construction standards based on DIN EN 13670 [44], (non-exhaustive schematic diagram)

*Beton-Kalender 2013* contains a contribution [53] that deals in detail with the topic of re-analysing existing concrete bridges.

## 2.8 Documents and assistance for practical applications

Besides the information given here, there are already publications available [13–19], backed up by numerous presentations, which explain the DAfStb guideline and provide support for practical applications.

In addition to the research reports [7, 9–11, 54, 55] and university publications [56, 57] written within the scope of drafting the DAfStb guideline, there is also a commentary available that explains the guideline section by section [58, 59]. It is therefore possible to follow the background to every provision in the guideline. This publication also includes many design examples in order to illustrate the use of the guideline in practice. The commentary with the design examples is also available in English [59].

# 3 Design of strengthening measures with externally bonded CFRP strips

## 3.1 Principles

The design of a strengthening measure must be carried out in such a way that the potential modes of failure are avoided. The various failure modes and their critical loads must be known in order to carry out the design with a margin corresponding to the level of safety for the critical load of the respective governing mode of failure. Figure 3.1 provides an overview of the failure modes that can occur. First of all, modes of failure related to the function of the CFRP strip can be added to those familiar from conventional reinforced concrete:

- Failure of concrete in compression zone
- Yielding of internal reinforcement followed by concrete failure
- Yielding of internal reinforcement followed by failure of the adhesively bonded strip
- Shear failure
- Yielding of externally bonded steel plate.

Besides these modes of failure well known from conventional reinforced concrete and relatively easy to describe, there are other modes specific to strengthening measures with externally bonded reinforcement. The first of these that should be mentioned is concrete cover separation failure, where the concrete cover becomes detached at the end of a strip. This occurs due to the additional, vertical offset between shear link and strip because the tensile stresses from the strip cannot be tied back to the compression zone of the beam. This mode of failure therefore corresponds to a horizontal shear failure in the area between the externally bonded reinforcement and the internal reinforcement.

The bond between the adhesive and the concrete often fails when using externally bonded reinforcement. In such a bond failure the layers of concrete near the surface break away once the tensile strength of the concrete has been exceeded. Owing to the only moderate tensile strength of the layers of concrete near the surface, following local debonding of externally bonded reinforcement, the result is mostly a total failure of the bond between the externally bonded reinforcement and the concrete as the load rises further because the forces involved cannot normally be carried by any remaining areas of intact bonding (unzipping effect). This behaviour means it is necessary to consider the bond of externally bonded reinforcement very carefully.

As flexural strengthening with externally bonded CFRP strips represents the most common form of strengthening and considerable research into this form of strengthening has been carried out in recent years, a staged analysis concept is available. We can choose between elaborate and simple analyses depending on requirements regarding accuracy or economics. Figure 3.2 provides an overview of this staged concept. Basically, designing flexural strengthening with CFRP strips always requires the designer to carry out a flexural analysis, in which failure of the concrete in compression or the reinforcement in tension is ruled out, a bond analysis and a shear analysis, and also to check for the risk of concrete cover separation failure.

*Strengthening of Concrete Structures with Adhesively Bonded Reinforcement: Design and Dimensioning of CFRP Laminates and Steel Plates*. First Edition. Konrad Zilch, Roland Niedermeier, and Wolfgang Finckh.

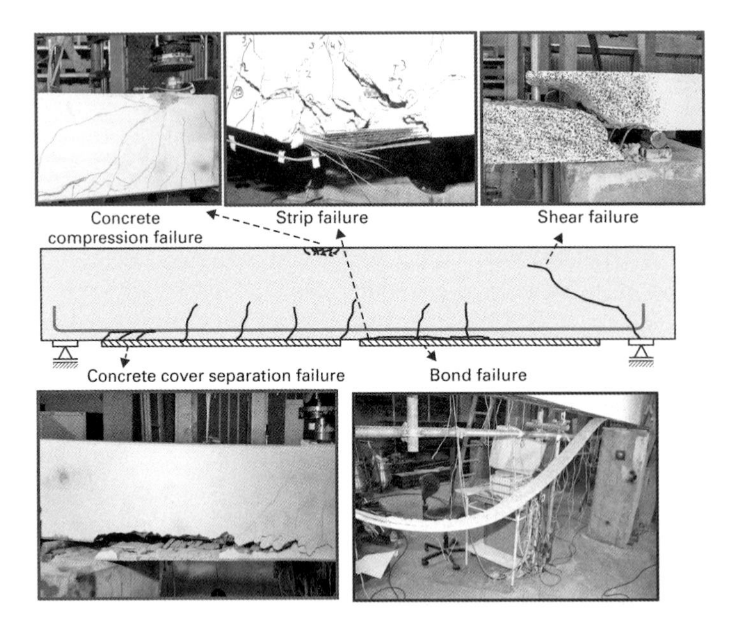


**Fig. 3.1** Examples of failure modes for a reinforced concrete beam with flexural strengthening in the form of externally bonded reinforcement

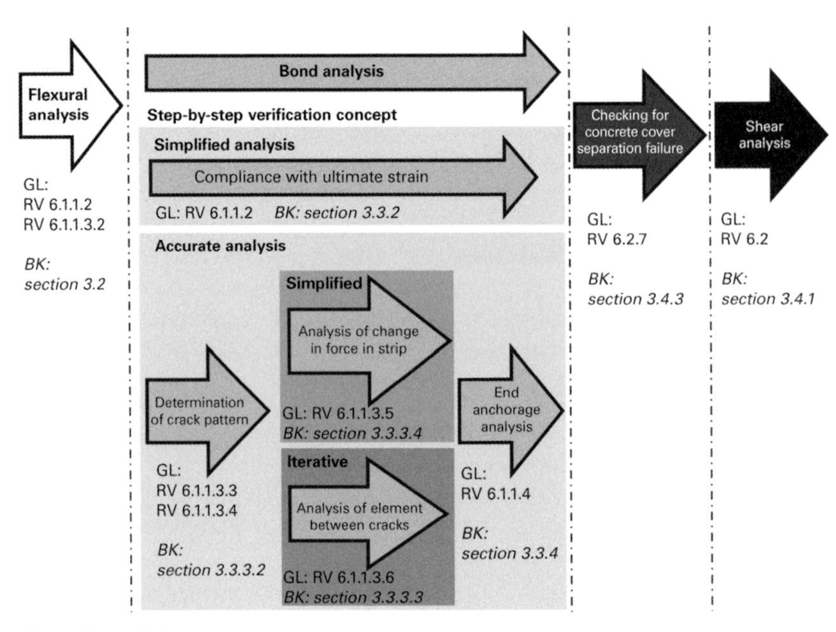


**Fig. 3.2** Flow chart for analysing flexural strengthening with CFRP strips

The bond analysis is complex and can be carried out in several stages depending on the accuracy requirements. An analysis of the end anchorage of the strip is required at every stage. In the simplified approach, only the ultimate strain in the externally bonded reinforcement has to be checked in addition to performing the end anchorage analysis. In contrast to this, in the more accurate method, the spacing of flexural cracks at the ultimate limit state must first be estimated. Either a simplified analysis of the change in the force in the strip is then carried out or an accurate iterative analysis of each concrete element between cracks. Figure 3.2 includes the section numbers of the DAfStb guideline [1, 2] and the section numbers of this book for the individual steps in the analysis.

## 3.2 Verification of flexural strength

The analysis of the flexural strength can be carried out in a similar way to that for a conventional reinforced concrete member by investigating the cracked cross-section. Equations 3.1 and 3.2 can be used to determine the forces via the equilibrium of the internal and external forces:

$$\sum M = 0: \quad M_{\text{Rd}} = M_{\text{Ed}} \tag{3.1}$$

$$\sum N = 0: \quad N_{\text{Rd}} = N_{\text{Ed}} \tag{3.2}$$

However, when determining the resistances, the function of the strip, and also any potential prestrain in the cross-section due to the loads during strengthening, must be taken into account. To do this, the equations known from conventional reinforced concrete (see [60–62], for example) must be extended, as it has been carried out, for instance, in [26] and also in the annex to the DAfStb guideline [1, 2]. In the following, these equations are specified in the same way as they are used in the examples in Sections 3.4 and 3.6. The cross-section is enlarged by the addition of the CFRP strip according to Figure 3.3 and so its resistance is expressed by Equations 3.3 and 3.4:

$$N_{\text{Rd}} = F_{\text{cd}} + F_{\text{Ld}} + F_{\text{s1d}} + F_{\text{s2d}} \tag{3.3}$$

$$M_{\text{Rd}} = -F_{\text{cd}} \cdot (z - z_{\text{L}}) + F_{\text{Ld}} \cdot z_{\text{L}} + F_{\text{s1d}} \cdot z_{\text{s1}} - F_{\text{s2d}} \cdot z_{\text{s2}} \tag{3.4}$$

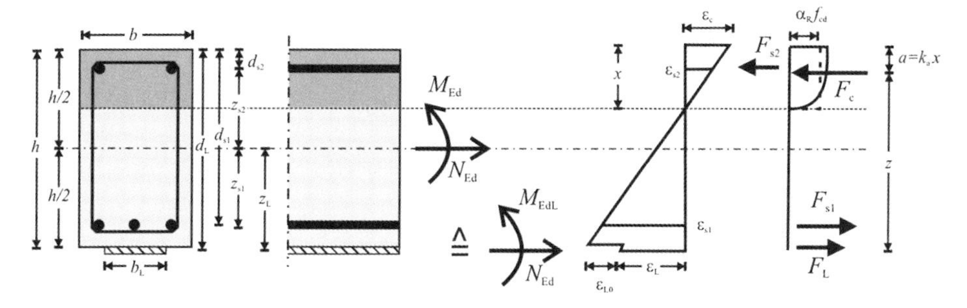


**Fig. 3.3** Geometry of strain distribution and member resistances of a strengthened and preloaded reinforced concrete cross-section

Translating the moment of resistance of the cross-section to the axis of the force in the strip results in Equation 3.5 and translating it to the axis of the compressive force in the concrete results in Equation 3.6:

$$\begin{aligned} M_{RdL} &= M_{Rd} - N_{Rd} \cdot z_L \\ &= -F_{cd} \cdot z - F_{s1d} \cdot (d_L - d_{s1}) - F_{s2d} \cdot (d_L - d_{s2}) \end{aligned} \tag{3.5}$$

$$\begin{aligned} M_{Rdc} &= M_{Rd} + N_{Rd} \cdot (z - z_L) \\ &= F_{Ld} \cdot (d_L - k_a \cdot x) + F_{s1d} \cdot (d_{s1} - k_a \cdot x) + F_{s2d} \cdot (k_a \cdot x - d_{s2}) \end{aligned} \tag{3.6}$$

In the above, coefficient $k_a$, which describes the magnitude of the compressive force in the concrete, is calculated depending on the compressive strain in the concrete $\varepsilon_c$ using Equation 3.7:

$$k_a = \begin{cases} \dfrac{8 + \varepsilon_c}{24 + 4 \cdot \varepsilon_c} & \text{for } \varepsilon_c \geq -2\ \text{mm/m} \\ \dfrac{3 \cdot \varepsilon_c^2 + 4 \cdot \varepsilon_c + 2}{6 \cdot \varepsilon_c^2 + 4 \cdot \varepsilon_c)} & \text{for } -2\ \text{mm/m} > \varepsilon_c \geq -3.5\ \text{mm/m} \end{cases} \tag{3.7}$$

The depth of the compression zone can be determined depending on the strain in the strip $\varepsilon_L$ and the compressive strain in the concrete $\varepsilon_c$ by taking into account the prestrain $\varepsilon_{L,0}$ using Equation 3.8:

$$x = \frac{-\varepsilon_c}{-\varepsilon_c + \varepsilon_{L,0} + \varepsilon_L} \cdot d_L \tag{3.8}$$

The internal compression in the concrete can be described via the compressive strength $f_{cd}$, the depth of the compression zone $x$, the width of the compression zone $b$ and the stress block geometry factor $\alpha_R$ using Equation 3.9. In this equation factor $\alpha_R$ takes into account the ratio of the mean concrete compressive stress to the concrete compressive strength and can be determined depending on the compressive strain in the concrete $\varepsilon_c$ using Equation 3.10:

$$F_{cd} = b \cdot x \cdot f_{cd} \cdot \alpha_R \tag{3.9}$$

$$\alpha_R = \begin{cases} \dfrac{-\varepsilon_c}{2} - \dfrac{\varepsilon_c^2}{12} & \text{for } \varepsilon_c \geq -2\ \text{mm/m} \\ 1 + \dfrac{2}{3 \cdot \varepsilon_c} & \text{for } -2\ \text{mm/m} > \varepsilon_c \geq -3.5\ \text{mm/m} \end{cases} \tag{3.10}$$

The tensile force in the strip is described via the strain in the strip, the modulus of elasticity and the area of the strip using Equation 3.11. The ultimate force for the strip may not be exceeded.

$$F_{Ld} = A_L \cdot E_L \cdot \varepsilon_L \leq A_L \cdot f_{Lud} \tag{3.11}$$

Equations 3.12 and 3.13 can be used to calculate the forces in the reinforcing steel in a similar way to the force in the strip. However, a force greater than the yield force may not be assumed. Strain hardening of the reinforcing steel in the plastic zone is neglected here.

$$F_{\mathrm{s1d}} = A_{\mathrm{s1}} \cdot E_{\mathrm{s}} \cdot \varepsilon_{\mathrm{s1}} \leq A_{\mathrm{s1}} \cdot f_{\mathrm{yd}} \tag{3.12}$$

$$F_{\mathrm{s2d}} = A_{\mathrm{s2}} \cdot E_{\mathrm{s}} \cdot \varepsilon_{\mathrm{s2}} \leq A_{\mathrm{s2}} \cdot f_{\mathrm{yd}} \tag{3.13}$$

Equations 3.14 and 3.15 can be used to determine the strains needed to calculate the forces in the reinforcing steel via the depth of the compression zone and the compressive strain in the concrete:

$$\varepsilon_{\mathrm{s1}} = -\varepsilon_{\mathrm{c}} \cdot \frac{d_{\mathrm{s1}} - x}{x} \tag{3.14}$$

$$\varepsilon_{\mathrm{s2}} = -\varepsilon_{\mathrm{c}} \cdot \frac{d_{s2} - x}{x} \tag{3.15}$$

With the help of the preceding equations it is possible to determine the strain in the strip and the compressive strain in the concrete iteratively via Equations 3.1 and 3.2. In doing so, the compressive strain in the concrete $\varepsilon_c$ may not drop below the value $\varepsilon_{cu2}$ to DIN EN 1992-1-1 [20] and the strain in the strip $\varepsilon_L$ may not exceed the ultimate strain $\varepsilon_{Lud}$. As in the normal case the ultimate strain $\varepsilon_{Lud}$ of the strip is significantly lower than the maximum strain in the reinforcing steel $\varepsilon_{su}$ to DIN EN 1992-1-1 [20], this limit for the reinforcing steel in the strengthened cross-section is not normally critical.

## 3.3 Bond analysis

### 3.3.1 Principles

As was already mentioned in Section 3.1, special considerations apply to the bond of externally bonded reinforcement. In conventional reinforced concrete construction, a bond analysis normally involves checking the end anchorage, which is based on bond values obtained from pull-out tests. If we carry out such an analysis in a similar form, then in members with externally bonded reinforcement the full tensile forces cannot be anchored because beyond a certain anchorage length it is not possible to increase the bond force substantially (see Figure 3.4). However, tests on flexural members have shown that much higher strip forces are reached at the point of maximum moment than would be possible via the end anchorage alone. For CFRP strips in particular, which can accommodate very high tensile stresses, only considering the end anchorage analysis would therefore be extremely uneconomic. The transfer of the bond force must instead take place at the point at which the changes in the tensile force occur, as indicated in Figure 3.4. For this reason, we distinguish between two areas when performing an analysis: the end anchorage region and the rest of the member. The strip forces at the flexural crack nearest the point of contraflexure must be anchored at the end anchorage point. The bond forces that can be accommodated in the end anchorage zone are determined by so-called idealized end anchorage tests in which the externally bonded reinforcement is peeled off in the longitudinal direction.

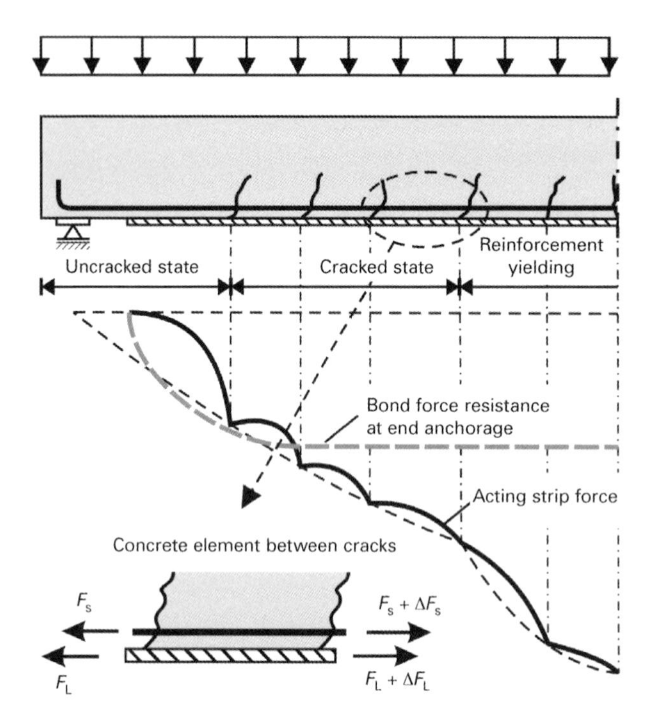


**Fig. 3.4** Principle of bond force transfer with externally bonded CFRP strips

For the rest of the member, the bond force can be transferred by concrete elements separated by flexural cracks. The forces prevailing in such a concrete element between cracks are a basic strip force at the less heavily stressed crack edge and this basic strip force plus an additional strip force at the more highly stressed crack edge. This additional strip force must be transferred into the member via bond.

### 3.3.2 Simplified method

In the simplified method it is only necessary to use Equation 3.16 to check that the ultimate strain of the strip is not exceeded:

$$\varepsilon_{\mathrm{Ld,max}} = \max\begin{cases} 0.5\,\mathrm{mm/m} + 0.1\,\mathrm{mm/m}\cdot\dfrac{l_0}{h} - 0.04\,\mathrm{mm/m}\cdot\phi_s + 0.06\,\mathrm{mm/m}\cdot f_{\mathrm{cm}} \\ \begin{cases} 3.0\,\mathrm{mm/m}\cdot\dfrac{l_0}{9700\,\mathrm{mm}}\cdot\left(2-\dfrac{l_0}{9700\,\mathrm{mm}}\right) & \text{for} \quad l_0 \le 9700\,\mathrm{mm} \\ 3.0\,\mathrm{mm/m} & \text{for} \quad l_0 > 9700\,\mathrm{mm} \end{cases} \end{cases} \quad (3.16)$$

This simplified analysis is based on a parametric study in [57] covering the more accurate analysis of the change in force in the strip at the concrete element between cracks in section RV 6.1.1.3.6 of the DAfStb guideline [1, 2], or Section 3.3.3.3 of this book, plus the analysis of the end anchorage at the final element between cracks. The parametric study assumed certain boundary conditions:

- The strip continues to within 50 mm from the front edge of the support.
- The internal reinforcing steel is ribbed.
- The internal reinforcing steel is not curtailed.
- The tensile strength of the concrete correlates with the compressive strength.
- The member is not prestressed.
- The strengthening is provided for positive moments (span moments).

With uniformly distributed loads it can be assumed that in the event that the first three points above are not fulfilled, then with an additional check of the end anchorage and the associated checking of the initial increase in the strip tensile force envelope, the simplified analysis also lies on the safe side. The correlation between the tensile and compressive strength of the concrete mentioned in the fourth point is achieved by adapting the concrete compressive strength in Equation 3.16 to the near-surface tensile strength in the DAfStb guideline using Equation 3.17:

$$f_{\text{ctm,surf}} \geq 0.26 \cdot f_{\text{cm}}^{2/3} \tag{3.17}$$

Strengthening for prestressed members cannot be designed with the simplified approach because in some circumstances the prestressing can change the distribution of the strains over the cross-section significantly. Likewise, strengthening in hogging moment zones, which occur in continuous beams, for example, cannot be designed with this method because there is an unfavourable relationship between moment and shear force. In addition, this method assumes that the member is cracked at the ultimate limit state.

### 3.3.3 More accurate method

#### 3.3.3.1 General

The more accurate method is based on the transfer of the bond force at the concrete element between cracks, as has already been briefly presented in Section 3.3.1 and Figure 3.4.

*Niedermeier* [63] was the first to formulate fundamental ideas about this, and he specifies a theoretical solution to this based on the differential equation of the bond-slip in [64]. Shortly afterwards, Neubauer [65] presented a solution for the bond force transfer at the element between cracks which is equivalent in terms of its mechanics.

To consider the transfer of the bond force, or rather the decrease in the tensile force, across the elements between cracks, the member is subdivided into several elements by means of the flexural cracks. It is necessary to distinguish between two areas here (see [65, 66]): the end anchorage region and the rest of the member. The strip forces at the flexural crack closest to the support must be anchored at the end anchorage point. The bond forces that can be accommodated in the end anchorage zone are determined by so-called idealized end anchorage tests in which the externally bonded reinforcement is peeled off in the longitudinal direction.

For the rest of the member, the bond force can be transferred to concrete elements separated by flexural cracks. The forces prevailing in such a concrete element between cracks are a basic strip force at the less heavily stressed crack edge and this basic strip force plus an additional strip force at the more highly stressed crack edge. This additional strip force must be transferred into the member via bond.

It is possible to solve the differential equation of the bond-slip with the boundary conditions of the element between cracks on the basis of the bilinear bond stress–slip relationship determined from the end anchorage tests (see [66, 67]). *Niedermeier* [66] and *Neubauer* [65] reach somewhat different expressions for this but these can be converted into each other, as has been shown in [7] and [67].

*Niedermeier's* [66] bond force transfer at the element between cracks was extended by Finckh [9, 57] on the basis of member-specific effects. This extension, which is reported in DAfStb publication 592 [9] (piecewise also published in English [68–71]), has essentially been incorporated in the DAfStb guideline.

Owing to their mechanics-based derivation, the expressions for the bond analyses given in the guideline are all dependent on the bond coefficients of the extended bilinear bond stress–slip relationship according to Figure 3.5.

The guideline specifies recommended values for the extended bilinear bond stress–slip relationship for externally bonded CFRP strips. These are based on an evaluation undertaken in [9] regarding common CFRP strip strengthening systems currently in use. The recommended values are given in Equations 3.18 to 3.20. The values also take into account the influence of the long-term durability of the concrete by way of the coefficients $\alpha_{cc}$ and $\alpha_{ct}$ to DIN EN 1992-1-1 [20] plus its associated National Annex [21]:

$$\tau_{L1k} = 0.366 \cdot \sqrt{\alpha_{cc} \cdot f_{cm} \cdot \alpha_{ct} \cdot f_{ctm,surf}} \tag{3.18}$$

$$s_{L0k} = 0.201 \text{ mm} \tag{3.19}$$

$$\tau_{LFk} = 10.8 \cdot \alpha_{cc} \cdot f_{cm}^{-0.89} \tag{3.20}$$

These bond values represent characteristic values in the sense of a 5% fractile. The evaluation in the test reports was carried out depending on the mean values of the input variables. In order to achieve the same circumstances on the building site, the input

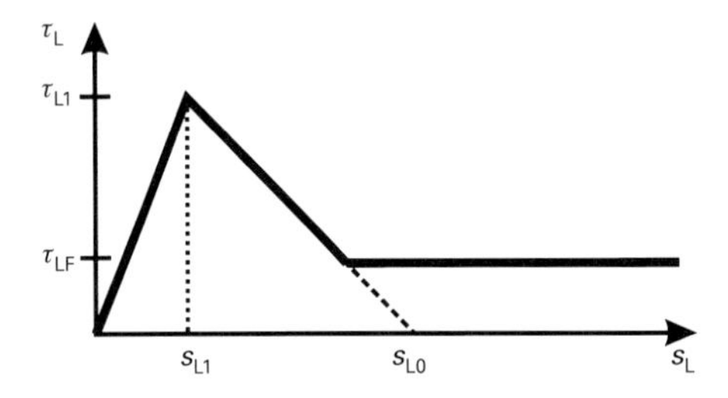


**Fig. 3.5** Extended bilinear bond stress–slip relationship

values must likewise be backed up by statistics. One key figure here is the near-surface tensile strength, which according to DIN EN 1542 [72] should always be determined on the member and according to the fourth part of the DAfStb guideline [1, 2] must be evaluated statistically, as it was also the case in DIN 1048-2 [73].

These values, given in annex K of the DAfStb guideline, only represent recommendations and may be dealt with differently in the national technical approvals for the systems. With strengthening systems that differ considerably from the CFRP strip systems approved hitherto, in some circumstances it can happen that the values are different to those recommended in the guideline.

#### 3.3.3.2 Determining the crack spacing

The more accurate bond analyses of the DAfStb guideline [1, 2] are dependent on the crack spacing, as Figure 3.4 shows. The formation of cracks in a strengthened reinforced concrete beam depends on many influences and is subject to considerable scatter. For this reason, the estimate of the crack spacing is approximated on the safe side in the DAfStb guideline. A simplified way of calculating the mean crack spacing for a stabilized crack pattern is to assume it is 1.5 times the transmission length of the reinforcing steel:

$$s_r = 1.5 \cdot l_{e,0} \tag{3.21}$$

The transmission length of the reinforcing steel is determined based on Noakowski [74] with Equation 3.22 using the cracking moment from Equation 3.23 and the mean bond force from Equation 3.24.

$$l_{e,0} = \frac{M_{cr}}{z_s \cdot F_{bsm}} \tag{3.22}$$

As the near-surface tensile strength has to be determined precisely for every member that is to be strengthened with externally bonded reinforcement, the cracking moment should also be derived from the parameters governing the member. Correlating the flexural tensile strength and the axial tensile strength can be carried out in numerous ways, with inputs including maximum aggregate diameter, member depth, etc. One generally acknowledged correlation method included in the DAfStb guideline [1, 2] is the relationship specified in DIN EN 1992-1-1:

$$M_{cr} = \kappa_{fl} \cdot f_{ctm,surf} \cdot W_{c,o} \tag{3.23}$$

where:

$\kappa_{fl} = (1.6 - h/1000) \geq 1.0$
$h$ total depth of member in mm

The mean bond force is determined here via the circumference of the reinforcing steel and the mean bond coefficient:

$$F_{bsm} = \sum_{i=1}^{n} n_{s,i} \cdot \phi_i \cdot \pi \cdot f_{bSm} \tag{3.24}$$

The values from Equation 3.25 are used for the mean bond stress in the reinforcing bars and depend on the type of bar used. These values are based on DIN EN 1992-1-1 [20] for ribbed reinforcing bars and the values from a simplified version of the approach by Noakowski [74] for plain reinforcing bars:

$$f_{bSm} = \begin{cases} \kappa_{vb1} \cdot 0.43 \cdot f_{cm}^{2/3} & \text{for ribbed rebars} \\ \kappa_{vb2} \cdot 0.28 \cdot \sqrt{f_{cm}} & \text{for plain rebars} \end{cases} \quad (3.25)$$

DAfStb publication 594 [11] states that solely considering the reinforcing steel for the crack widths results in the crack spacing being overestimated because with surface-mounted CFRP strips as well, the crack spacing is influenced by the composite action of the concrete. The various bond stiffnesses of and strains in the lines of reinforcement must be included if we are to achieve a more accurate calculation of the crack spacing, as was done in [57, 75], for instance. However, such an approach increases the amount of calculation required because the ensuing crack spacings depend on the area of the surface-mounted reinforcement, which is not known ahead of the design work. The larger crack spacings determined by neglecting the effect of the surface-mounted reinforcement lead, however, to results that lie on the safe side, which permit this simplification.

#### 3.3.3.3 Accurate analysis of concrete element between cracks

The accurate analysis of the concrete element between cracks involves checking that in the cracked area of the member the change in force in the strip in the element between cracks $\Delta F_{LEd}$, which is characterized by the shear force, is smaller than the change in force that can be accommodated by bond (see also Figure 3.4).

$$\Delta F_{LEd} \leq \Delta F_{LRd} \quad (3.26)$$

$$\Delta F_{LEd} = F_{LEd}(x + s_r) - F_{LEd}(x) \quad (3.27)$$

One of the things on which the admissible change in the strip force in the element between cracks depends is the strip force at the less heavily stressed crack edge $F_{LEd}(x)$, which is characterized by the bending moment. As was described in [76] for the first time, the admissible change in strip force in the element between cracks is divided into three effects: the component from the bilinear bond stress–slip relationship $\Delta F_{Lk,BL}$ according to [66, 77], the component from an additional frictional bond that occurs at the places where debonding has already taken place $\Delta F_{Lk,BF}$ according to [9] and the component from curvature $\Delta F_{Lk,KF}$ according to [9]:

$$\Delta F_{LRd} = \frac{\Delta F_{Lk,BL} + \Delta F_{Lk,BF} + \Delta F_{Lk,KF}}{\gamma_{BA}} \quad (3.28)$$

The three components are shown schematically in Figure 3.6 via the strip force at the less heavily stressed crack edge of the element between cracks together with the associated bond stress–slip relationships. The equations for describing the individual components are given below and briefly explained; a full description plus the derivation

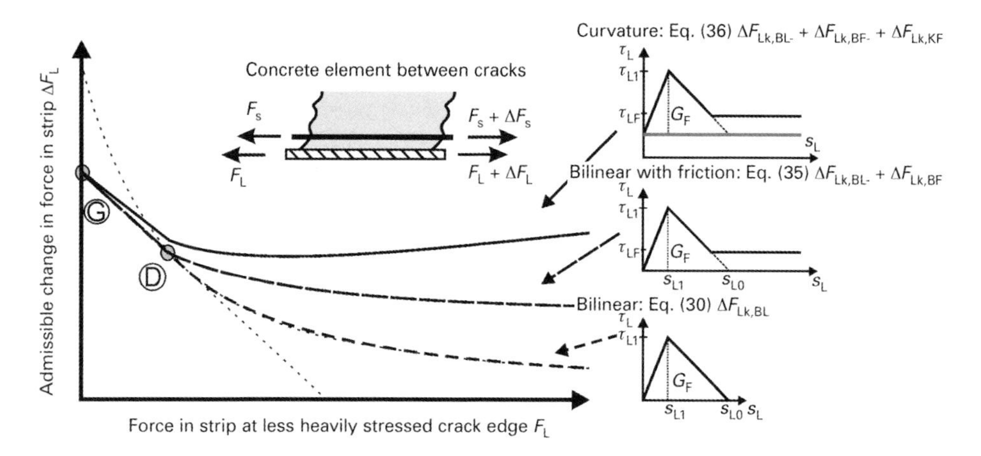


**Fig. 3.6** Change in force in strip that can be accommodated in concrete element between cracks due to the three components depending on the force in the strip at the less heavily stressed crack edge

of the equations can be found in DAfStb publication 592 [9]. The equations depend on the geometric variables of the bonded reinforcement (width $b_L$, theoretical thickness $t_L$), the material properties of the bonded reinforcement (design value of ultimate strength $F_{Lud}$, mean modulus of elasticity $E_{Lm}$) and the bond coefficients of the extended bilinear bond stress–slip relationship, and hence on the maximum bond stress $\tau_{L1}$, the maximum slip $s_{L0}$ and the frictional bond stress $\tau_{LF}$ (see Section 3.3.3.1).

The first component in Equation 3.28, which describes the bond strength from the bilinear bond stress–slip relationship at the element between cracks according to *Niedermeier* [66, 77], is divided into two parts by point D in Figure 3.6 and can be determined with Equation 3.29. The first part, from point G to point D, described by a straight line between these two points, represents the range over which the required transfer length of the bilinear bond stress–slip model is greater than the length of the element between cracks $s_r$.

$$\Delta F_{\mathrm{Lk,BL}} = \begin{cases} \Delta F^{\mathrm{G}}_{\mathrm{Lk,BL}} - \dfrac{\Delta F^{\mathrm{G}}_{\mathrm{Lk,BL}} - \Delta F^{\mathrm{D}}_{\mathrm{Lk,BL}}}{F^{\mathrm{D}}_{\mathrm{Lk,BL}}} F_{\mathrm{LEd}} & \text{for} \quad F_{\mathrm{LEd}} \leq F^{\mathrm{D}}_{\mathrm{Lk,BL}} \\ \sqrt{b_{\mathrm{L}}^2 \cdot \tau_{\mathrm{L1k}} \cdot s_{\mathrm{L0k}} \cdot E_{\mathrm{Lm}} \cdot t_L + F^2_{\mathrm{LEd}}} - F_{\mathrm{LEd}} & \text{for} \quad F^{\mathrm{D}}_{\mathrm{Lk,BL}} < F_{\mathrm{LEd}} \leq F_{\mathrm{Lud}} \end{cases} \tag{3.29}$$

The forces required for points G and D are calculated with Equations 3.30 to 3.32. The effective bond length $l_{bL,max}$ required for this can be determined via the bond parameters of the bilinear bond stress–slip relationship and the empirical calibration coefficient $\kappa_{Lb} = 1.128$ according to *Niedermeier* [66] using Equation 3.33.

$$\Delta F_{\mathrm{Lk,BL}}^{\mathrm{G}} = \begin{cases} b_{\mathrm{L}}\sqrt{E_{\mathrm{Lm}} \cdot s_{\mathrm{L0k}} \cdot \tau_{\mathrm{L1k}}} \cdot \dfrac{s_{\mathrm{r}}}{l_{\mathrm{bL,max}}}\left(2 - \dfrac{s_{\mathrm{r}}}{l_{\mathrm{bL,max}}}\right) & s_{\mathrm{r}} < l_{\mathrm{bL,max}} \\ b_{\mathrm{L}}\sqrt{E_{\mathrm{Lm}} \cdot s_{\mathrm{L0k}} \cdot \tau_{\mathrm{L1k}}} & s_{\mathrm{r}} \geq l_{\mathrm{bL,max}} \end{cases} \tag{3.30}$$

$$F_{\mathrm{Lk,BL}}^{\mathrm{D}} = \frac{s_{\mathrm{L0k}} \cdot E_{\mathrm{Lm}} \cdot b_{\mathrm{L}} \cdot t_{\mathrm{L}}}{s_{\mathrm{r}}} - \tau_{\mathrm{L1k}} \frac{s_{\mathrm{r}} b_{\mathrm{L}}}{4} \tag{3.31}$$

$$\Delta F_{\mathrm{Lk,BL}}^{\mathrm{D}} = \sqrt{b_L^2 \cdot \tau_{\mathrm{L1k}} \cdot s_{\mathrm{L0k}} \cdot E_{\mathrm{Lm}} \cdot t_{\mathrm{L}} + F_{\mathrm{Lk,BL}}^{\mathrm{D}}{}^2} - F_{\mathrm{Lk,BL}}^{\mathrm{D}} \tag{3.32}$$

$$l_{\mathrm{bL,max}} = \frac{2}{\kappa_{\mathrm{Lb}}} \cdot \sqrt{\frac{E_{\mathrm{Lm}} \cdot t_{\mathrm{L}} \cdot s_{\mathrm{L0k}}}{\tau_{\mathrm{L1k}}}} \tag{3.33}$$

The second component from the frictional bond between the surface already debonded, which can only occur after point D in Figure 3.6, is calculated in the DAfStb guideline according to DAfStb publication 592 [9] (q.v. [71]) using Equation 3.34:

$$\Delta F_{\mathrm{Lk,BF}} = \begin{cases} 0 \quad \text{for } F_{\mathrm{LEd}} \leq F_{\mathrm{Lk,BL}}^{\mathrm{D}} \\ \tau_{\mathrm{LFk}} \cdot b_{\mathrm{L}} \cdot \left(s_{\mathrm{r}} - \dfrac{2 \cdot t_{\mathrm{L}} \cdot E_{\mathrm{Lm}}}{\tau_{\mathrm{L1k}}} \cdot \left(\sqrt{\dfrac{\tau_{\mathrm{L1k}} \cdot s_{\mathrm{L0k}}}{t_{\mathrm{L}} \cdot E_{\mathrm{Lm}}} + \dfrac{F_{\mathrm{LEd}}^2}{b_{\mathrm{L}}^2 \cdot t_{\mathrm{L}}^2 \cdot E_{\mathrm{Lm}}^2}} - \dfrac{F_{\mathrm{LEd}}}{b_{\mathrm{L}} \cdot t_{\mathrm{L}} \cdot E_{\mathrm{Lm}}}\right)\right) \\ \qquad \text{for } F_{\mathrm{Lk,BL}}^{\mathrm{D}} < F_{\mathrm{LEd}} \leq F_{\mathrm{Lud}} \end{cases} \tag{3.34}$$

The third component in Equation 3.28 represents how the curvature of the member influences the bond of the surface-mounted reinforcement. *Zilch et al.* [76] (q.v. [68]) were the first to investigate and quantify this effect. A convex curvature, as caused by deflection, causes a change in direction at each concrete element between cracks, which therefore leads to a self-induced contact pressure. This contact pressure on the surface-mounted reinforcement brings about an increase in the bond strength. In the DAfStb guideline this effect is expressed via the curvature of the cross-section using Equation 3.35, a simplified expression that uses the depth of the member $h$, the compressive strain in the concrete $\varepsilon_{\mathrm{cr1}}$ and the strain in the strip $\varepsilon_{\mathrm{Lr1}}$. Equation 3.35 includes the empirical coefficient $\kappa_{\mathrm{k}} = 24.3 \times 10^3$ N/mm to take into account the influence of the curvature on the bond, which was determined by means of numerous tests in DAfStb publication 592 [9] (q.v. [70]):

$$\Delta F_{\mathrm{Lk,KF}} = s_{\mathrm{r}} \cdot \kappa_{\mathrm{k}} \cdot \frac{\varepsilon_{\mathrm{Lr1}} - \varepsilon_{\mathrm{cr1}}}{h} \cdot b_{\mathrm{L}} \tag{3.35}$$

The accurate analysis of the concrete element between cracks presented here tends to be unsuitable for manual calculations because the critical point for the design is not readily discernible, instead first appears at the end of the entire analysis. If we consider, for example, a two-span beam subjected to a uniformly distributed load, then it takes considerable effort to determine the critical load case for the most unfavourable combination at the critical element between cracks, which tends to make this analysis impractical for manual calculations. However, this analysis is relatively well suited to computer calculations because case distinction is hardly necessary.

#### 3.3.3.4 Simplified analysis of element between cracks

A simpler option given in the DAfStb guideline [1, 2] for analysing the bond of the externally bonded reinforcement is to limit the change in the force in the strip, as is shown schematically in Figure 3.7.

In this analysis it is only necessary to verify that the change in the strip force does not exceed a constant resistance value at any point in the member. This resistance value, represented by the dotted line in Figure 3.7, was determined via a numerical approach to the more accurate method for the limits prescribed in DAfStb publication 592 [9] (q.v. [70]), the background to this being to separate resistance from action in order to be able to use the superposition principle again. The proposal for this analysis in the DAfStb guideline is Equation 3.36, which again depends on the bond coefficients and the length of the element between cracks. In addition, the equation depends on the

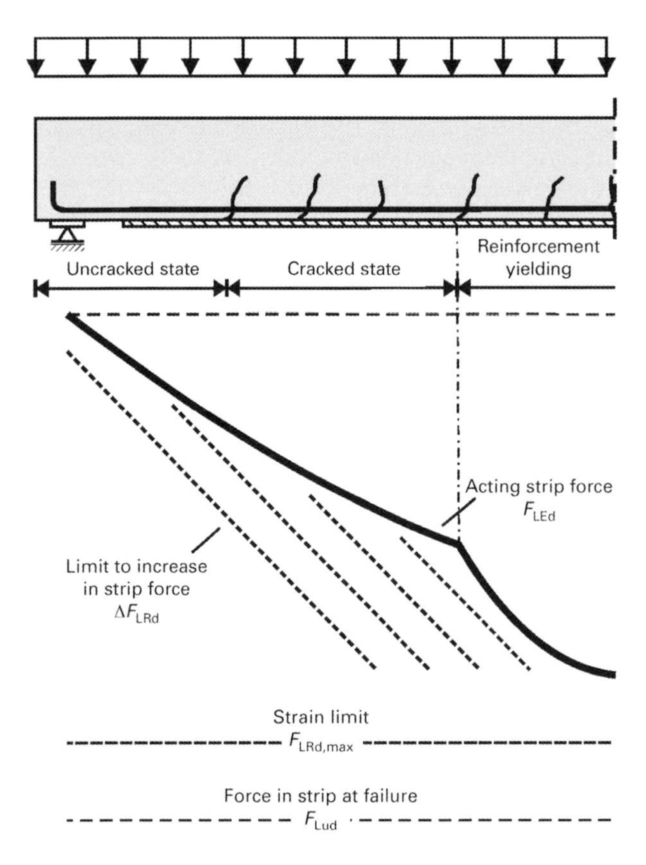


**Fig. 3.7** Scheme for analyses to be carried out for ultimate strain in strip and change in force in strip in concrete element between cracks (simplified analysis)

factor $\kappa_h = 2000$ for planar reinforced concrete members and $\kappa_h = 0$ for prestressed concrete members, which likewise take the influence of curvature into account. As the proposed equation was worked out numerically between certain limits, these may not be exceeded. One of these limits is the additional ultimate strain for bonded reinforcement amounting to 10 mm/m, which may not be exceeded. It should be noted that Equation 3.36 has various units and all values must therefore be entered in N and mm.

$$\Delta F_{LRd} = \frac{\tau_{L1k} \cdot 2.3 \cdot \sqrt{s_r} + \tau_{LFk} \cdot 0.098 \cdot s_r^{4/3} + \frac{\kappa_h}{h} \cdot s_r^{1/3}}{\gamma_{BA}} \cdot b_L \tag{3.36}$$

### 3.3.4 End anchorage analysis

#### 3.3.4.1 General

As was already explained in Section 3.3.3.1, an analysis of the end anchorage is necessary in addition to checking the bond at the element between cracks. According to the DAfStb guideline [1, 2], the end anchorage analysis can be performed in three different ways (see Figure 3.8) depending on requirements.

#### 3.3.4.2 End anchorage analysis at flexural crack nearest to point of contraflexure

The analysis at the flexural crack closest to the point of contraflexure represents the standard case (Figure 3.8a). In this case the moment acting at this flexural crack must be lower than the resistance of the cross-section taking into account the 'shift rule' according to DIN EN 1992-1-1. The resistance of the cross-section is determined based

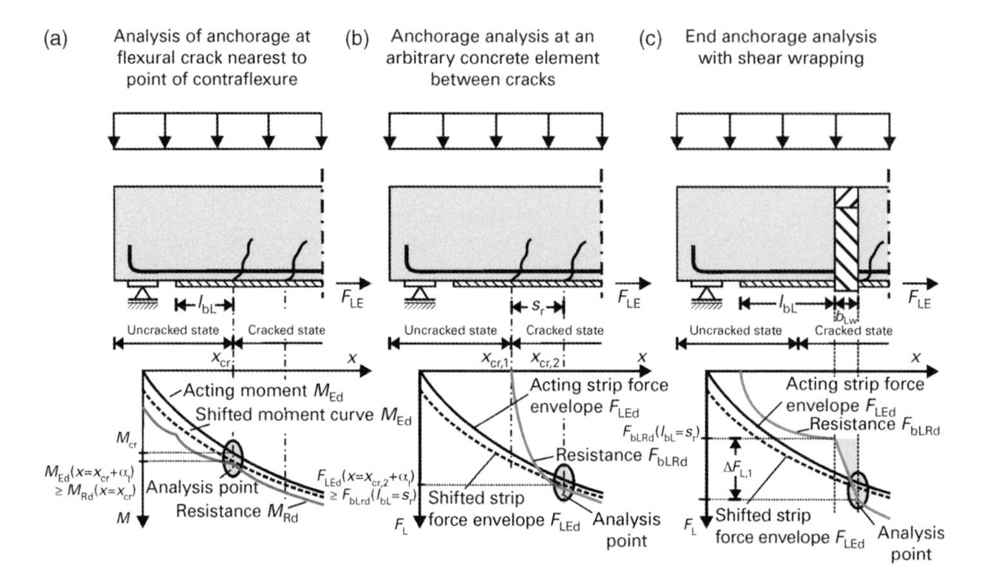


**Fig. 3.8** Scheme for analysing the three different options for verifying the end anchorage of externally bonded CFRP strips and CF sheets

on bond according to the concept of *Zehetmeier* [75, 78] (q.v. [79]), which considers a redistribution between the externally bonded and the internal reinforcement. Owing to the different bond behaviour and depending on the strain state of the bonded reinforcement, a different distribution of the forces between the various lines of reinforcement occurs, which is described via the slip of the strip.

The analysis is carried out at the position of the flexural crack nearest the point of contraflexure. As the analysis takes into account the interaction of the lines of reinforcement, it includes the acting moment and the moment that can be accommodated by the cross-section according to Equation 3.37:

$$M_{\mathrm{Ed}} \leq M_{\mathrm{Rd}}(l_{\mathrm{bL}}) \tag{3.37}$$

The admissible moment is determined depending on the strains in the lines of reinforcement using Equation 3.38. In doing so, a sufficiently long anchorage length is assumed for the reinforcing steel.

$$M_{\mathrm{Rd}}(l_{\mathrm{bL}}) = \varepsilon^{\mathrm{a}}_{\mathrm{LRk}}(l_{\mathrm{bL}}) \cdot E_{\mathrm{Lm}} \cdot A_{\mathrm{L}} \cdot z^{\mathrm{a}}_{\mathrm{L}} \cdot \frac{1}{\gamma_{\mathrm{BA}}} + \varepsilon^{\mathrm{a}}_{\mathrm{sRk}}(l_{\mathrm{bL}}) \cdot E_{\mathrm{s}} \cdot A_{\mathrm{s}} \cdot z^{\mathrm{a}}_{\mathrm{s}} \cdot \frac{1}{\gamma_{\mathrm{S}}} \tag{3.38}$$

Equation 3.39 is used to calculate the strain in the strengthening element depending on the bond length available beyond the flexural crack closest to the point of contraflexure. Here, the effective bond length $l_{\mathrm{bL,lim}}$ and the maximum strain $\varepsilon_{\mathrm{LRk,lim}}$ are calculated via the variables of the bilinear bond stress–slip relationship according to Section 3.3.3.1 using Equations 3.40 to 3.43 ($\kappa_{\mathrm{Lb}} = 1.128$, see Section 3.3.3.3).

$$\varepsilon^{\mathrm{a}}_{\mathrm{LRk}}(l_{\mathrm{bL}}) = \begin{cases} \sin\left(\dfrac{\pi}{2} \cdot \dfrac{l_{\mathrm{bL}}}{l_{\mathrm{bL,lim}}}\right) \cdot \varepsilon^{\mathrm{a}}_{\mathrm{LRk,lim}} & \text{for} \quad 0 < l_{\mathrm{bL}} < l_{\mathrm{bL,lim}} \\ \varepsilon^{\mathrm{a}}_{\mathrm{LRk,lim}} & \text{for} \quad l_{\mathrm{bL,lim}} \leq l_{\mathrm{bL}} \end{cases} \tag{3.39}$$

$$\varepsilon^{\mathrm{a}}_{\mathrm{LRk,lim}} = 0.985 \cdot \frac{f_{\mathrm{bLk,max}}}{E_{\mathrm{Lm}}} \tag{3.40}$$

$$f_{\mathrm{bLk,max}} = \sqrt{\frac{E_{\mathrm{Lm}} \cdot s_{\mathrm{L0k}} \cdot \tau_{\mathrm{L1k}}}{t_{\mathrm{L}}}} \tag{3.41}$$

$$l_{\mathrm{bL,lim}} = 0.86 \cdot l_{\mathrm{bL,max}} \tag{3.42}$$

$$l_{\mathrm{bL,max}} = \frac{2}{\kappa_{\mathrm{Lb}}} \cdot \sqrt{\frac{E_{\mathrm{Lm}} \cdot t_{\mathrm{L}} \cdot s_{\mathrm{L0k}}}{\tau_{\mathrm{L1k}}}} \tag{3.43}$$

The strains in the reinforcing steel are calculated depending on the slip of the strip $s_{\mathrm{Lr}}$, the bond factor $\kappa_{\mathrm{bsk}}$ and the weighting of the different lever arms according to Equation 3.43. Here, $\alpha_{\mathrm{N}} = 0.25$ for ribbed reinforcing bars and $\alpha_{\mathrm{N}} = 0$ for plain bars, and $\kappa_{\mathrm{VB}} = 1$ for good bond conditions and $\kappa_{\mathrm{VB}} = 0.7$ for moderate conditions. The bond factor $\kappa_{\mathrm{bsk}}$ is calculated according to Equation 3.45 using the values given in Table 3.1 according to [75].

**Table 3.1** Bond coefficients for internal reinforcement for the end anchorage analysis at the flexural crack nearest the point of contraflexure.

| Internal reinforcement | Ribbed | Plain |
|---|---|---|
| $\kappa_{b1k}$ | 2.545 | 1.292 |
| $\kappa_{b2}$ | 1.0 | 1.3 |
| $\kappa_{b3}$ | 0.8 | 1.0 |
| $\kappa_{b4}$ | 0.2 | 0.3 |

$$\varepsilon_{sRk}^{a}(l_{bL}) = \kappa_{VB} \cdot \kappa_{bsk} \cdot \left(s_{Lr}^{a}(l_{bL})\right)^{(\alpha_N+1)/2} \cdot \left(\frac{d^a - x^a}{d_L^a - x^a}\right)^{(\alpha_N+1)/2} \leq f_{yk}/E_s \tag{3.44}$$

$$s_{Lr}^{a}(l_{bL}) = \begin{cases} 0.213\,\text{mm} \cdot \left(1 - \cos\left(\frac{\pi}{2} \cdot \frac{l_{bL}}{l_{bL,lim}}\right)\right) & \text{for} \quad 0 < l_{bL} \leq l_{bL,lim} \\ 0.213\,\text{mm} + \left(l_{bL} - l_{bL,lim}\right) \cdot \varepsilon_{LRk,lim}^{a} & \text{for} \quad l_{bL,lim} \leq l_{bL} \end{cases} \tag{3.45}$$

$$\kappa_{bsk} = \kappa_{b1k} \cdot \sqrt{\frac{f_{cm}^{\kappa_{b2}}}{E_s \cdot \phi^{\kappa_{b3}} \cdot (E_{Lm} \cdot t_L)^{\kappa_{b4}}}} \tag{3.46}$$

#### 3.3.4.3 Anchorage analysis at an arbitrary concrete element between cracks

The second option – carrying out the end anchorage analysis at an arbitrary element between cracks – may be necessary for those members in which owing to the low tensile strength of the concrete, the flexural crack closest to the point of contraflexure is extremely close to the support. In this analysis the externally bonded reinforcement has to be anchored at an arbitrary element between cracks similarly to the analysis for the element between cracks according to Section 3.3.3.4 (Figure 3.8b). Besides taking into account the 'shift rule' in this analysis, it has to be ensured that the cross-section between the support and the element between cracks being considered possesses sufficient load-carrying capacity even without the externally bonded reinforcement.

As shown in Figure 3.8b, the last element between cracks must be checked to ensure that the acting strip force without redistribution is less than the bond resistance at the idealized end anchorage body:

$$F_{LEd} \leq F_{bLRd} \tag{3.47}$$

In this analysis the last element between cracks may be positioned at the end of the strip and may have a length corresponding to the crack spacing according to Section 3.3.3.2. The resistance to debonding at the last element between cracks represents a similar situation to that tested on the idealized end anchorage body and evaluated on the basis of the bilinear bond stress–slip relationship in [65, 80, 81], for example. However, in the analysis of the last concrete element between cracks, the bond length is limited to the crack spacing. Therefore, the resistance is given by Equations 3.48 to 3.52. The variables for the bilinear bond stress–slip relationship are listed in Section 3.3.3.1 and $\kappa_{Lb} = 1.128$ (see also Section 3.3.3.3).

$$F_{bLRd} = b_L \cdot t_L \cdot f_{bLd}(s_r) \tag{3.48}$$

$$f_{bLd}(s_r) = \frac{f_{bLk}(s_r)}{\gamma_{BA}} \tag{3.49}$$

$$f_{bLk}(s_r) = \begin{cases} f_{bLk,max} \cdot \dfrac{s_r}{l_{bL,max}}\left(2 - \dfrac{s_r}{l_{bL,max}}\right) & s_r < l_{bL,max} \\ f_{bLk,max} & s_r \geq l_{bL,max} \end{cases} \tag{3.50}$$

$$f_{bLk,max} = \sqrt{\frac{E_{Lm} \cdot s_{L0k} \cdot \tau_{L1k}}{t_L}} \tag{3.51}$$

$$l_{bL,max} = \frac{2}{\kappa_{Lb}} \cdot \sqrt{\frac{E_{Lm} \cdot t_L \cdot s_{L0k}}{\tau_{L1k}}} \tag{3.52}$$

#### 3.3.4.4 End anchorage analysis with shear wrapping

In the third option – analysis of end anchorage with shear wrapping – the shear wrapping, mostly provided as a result of the design for shear or to avoid a concrete cover separation failure, may also be employed to increase the bond force. The concept of increasing the bond force as a consequence of shear wrapping has been borrowed from *Husemann* [82] (q.v. [83]). To do this, as can be seen in Figure 3.8c, the resistance according to Equation 3.53 at the point at the end of the shear wrapping is compared with the shifted strip force envelope. In Equation 3.53 the increase in the bond force due to shear wrapping $\Delta F_{L,1}$ is added to the end anchorage force of the segment of strip beyond the shear wrapping, which is calculated similarly to Section 3.3.4.3:

$$F_{bLRd} = b_L \cdot t_L \cdot f_{bLd}(l_{bL}) + \frac{\Delta F_{L,1}}{\gamma_{BA}} \tag{3.53}$$

The increase in the bond force as a result of shear wrapping can be calculated with Equation 3.54 depending on the width of the strap $b_{Lw}$, the contact pressure $F_u$ ($\alpha_b$) and the factor $\kappa_l$ to take into account the form of the wrapped strip cross-section:

$$\Delta F_{L,1} = \frac{t_L \cdot b_L \cdot b_{Lw}}{120} \cdot \frac{\sqrt{f_{ctm,surf}}}{1.33} \cdot \left[230 \cdot \kappa_l \cdot \frac{F_u(\alpha_b)}{b_L \cdot b_{Lw}} - 23 \cdot \left(\kappa_l \cdot \frac{F_u(\alpha_b)}{b_L \cdot b_{Lw}}\right)^2\right] \tag{3.54}$$

To calculate the contact pressure, a distinction is made between the contact pressures $F_{u,2}$ and $F_{u,4}$, which are always formed by the two geometric limit cases shown in Figure 3.9. Interpolation between the two limit cases with the help of Equation 3.55 is

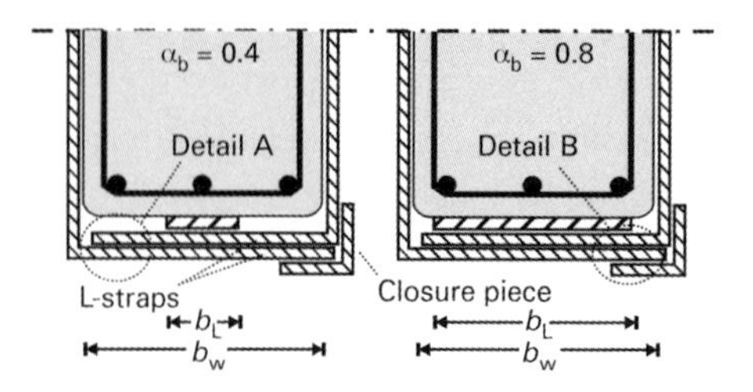


**Fig. 3.9** Section through downstand beam with externally bonded CFRP strip and shear wrapping

possible depending on the geometry factor $0.4 \leq \alpha_b = b_L/b_w \leq 0.8$:

$$F_u(\alpha_b) = F_{u,2} \cdot \left(\frac{0.8 - \alpha_b}{0.4}\right) + F_{u,4} \cdot \left(\frac{\alpha_b - 0.4}{0.4}\right) \tag{3.55}$$

The contact pressures depend on the stiffness of the shear wrapping in all cases. It is therefore necessary to calculate the stiffness of the shear wrapping first, which according to the detailing rules of Figure 3.9 is generally made up of two L-straps and one closure piece bonded with adhesive (see also Section 3.7). Consequently, we should distinguish between Detail A, consisting of two bonded L-straps, and Detail B, with two L-straps plus one closure piece (see Figure 3.9). Calculating the increase in the bond force must include determining the stiffness depending on the cross-sectional area $A_s$ and the moment of inertia $I_s$ of steel L-plates for both details. The stiffness for **Detail A** is given by Equations 3.56 and 3.57:

$$EI_{S,A} = 2 \cdot E_S \cdot \left(I_S + A_S \cdot z_{S,A}^2\right) \tag{3.56}$$

$$z_{S,A} = \frac{1}{2} \cdot t_{Lw} + 0.5 \tag{3.57}$$

Correspondingly, the stiffness for **Detail B** is given by Equations 3.58 and 3.59:

$$EI_{S,B} = 2 \cdot E_S \cdot \left(I_S + A_S \cdot z_{S,B}^2\right) + E_S I_S \tag{3.58}$$

$$z_{S,B} = t_{Lw} + 1 \tag{3.59}$$

Using these variables it is now possible to determine the contact pressures for the two limit cases $\alpha_b = 0.4$ and 0.8. The contact pressure for **limit case $\alpha_b = 0.4$** can be calculated using Equations 3.60 to 3.63. This results in the lengths $l_1 = 0.3 \cdot b_w - 20$ and $l_2 = b_w - 40$. The crack width for CFRP strips is then $w = 0.35$.

$$F_{u,2} = \frac{2 \cdot 24 \cdot EI_{s,g,\alpha_b=0.4}}{(3 \cdot \alpha - 4 \cdot \alpha^3) \cdot l_2^3} \cdot w_1 + \frac{26\,400 \cdot EI_{s,g,\alpha_b=0.4}}{11\,000 \cdot l_1^3 + 2.4 \cdot EI_{s,g,\alpha_b=0.4}} \tag{3.60}$$

$$w_1 = w - \left(1 - \frac{EI_{s,g,\alpha_b=0.4}}{4583 \cdot l_1^3 + EI_{s,g,\alpha_b=0.4}}\right) \cdot 0.1 \tag{3.61}$$

$$\alpha = \frac{0.3 \cdot b_{\text{w}} - 20}{b_{\text{w}} - 40} \tag{3.62}$$

$$EI_{\text{s,g},\alpha_{\text{b}} = 0.4} = 2 \cdot \frac{EI_{\text{S,A}} \cdot EI_{\text{S,B}}}{EI_{\text{S,A}} + EI_{\text{S,B}}} \tag{3.63}$$

Accordingly, the contact pressure for **limit case $\alpha_b = 0.8$** is given by Equations 3.64 to 3.66. This results in the lengths $l_3 = 20 + t_{\text{LW}}$ and $l_4 = 2 \cdot l_3$. The crack width for CFRP strips is then $w = 0.35$.

$$F_{\text{u},4} = \frac{48 \cdot EI_{\text{s,g},\alpha_{\text{b}} = 0.8}}{l_4^3} \cdot w_2 + \frac{26\,400 \cdot EI_{\text{s,g},\alpha_{\text{b}} = 0.8}}{11\,000 \cdot l_3^3 + 2.4 \cdot EI_{\text{s,g},\alpha_{\text{b}} = 0.8}} \tag{3.64}$$

$$w_2 = w - \left(1 - \frac{EI_{\text{s,g},\alpha_{\text{b}} = 0.8}}{4583 \cdot l_3^3 + EI_{\text{s,g},\alpha_{\text{b}} = 0.8}}\right) \cdot 0.1 \tag{3.65}$$

$$EI_{\text{s,g},\alpha_{\text{b}} = 0.8} = 2 \cdot \frac{EI_{\text{S,A}} \cdot E_{\text{S}} I_{\text{S}}}{EI_{\text{S,A}} + E_{\text{S}} I_{\text{S}}} \tag{3.66}$$

## 3.4 Shear force analyses

### 3.4.1 Shear strength

The DAfStb guideline [1, 2] states that the analyses of DIN EN 1992-1-1 [20] together with its National Annex [21] must be carried out to assess the shear strength. It has been shown in tests [11, 54] (q.v. [84]) that these analyses can also be used for strengthened members in the building stock. However, externally bonded reinforcement may not be counted as part of the longitudinal reinforcement ratio in Equation 6.2a of DIN EN 1992-1-1 [20].

In members with externally bonded flexural strengthening, debonding due to offset crack edges caused by the shear force can take place additionally in the case of high stresses in the tension and compression members of the truss assumed for carrying the shear force. For this reason, the guideline includes Equation 3.67, specifying a limit value above which the externally bonded CFRP strips have to be wrapped with externally bonded shear links:

$$\frac{V_{\text{Ed}} \cdot \sigma_{\text{sw}}}{V_{\text{Rd,max}}} \leq \begin{cases} 75\,\text{N/mm}^2 & \text{for ribbed shear links} \\ 25\,\text{N/mm}^2 & \text{for plain shear links} \end{cases} \tag{3.67}$$

The given limits are based on modelling and a subsequent parametric study in [57] (q.v. [85]). The shear link stress included in Equation 3.67 can be calculated by rearranging Equation 6.8 or 6.13 from DIN EN 1992-1-1 [20], as Equation 3.68 illustrates:

$$\sigma_{\text{sw}} = \frac{V_{\text{Ed}}}{(A_{\text{sw}}/s) \cdot z \cdot \cot\theta} \tag{3.68}$$

If the limit according to Equation 3.67 is exceeded, the force for the surface-mounted links required can be calculated via the ratio of the stiffnesses of the lines of longitudinal reinforcement:

$$V_{\mathrm{LEd}} = \max \begin{cases} \dfrac{EA_{\mathrm{L}}}{EA_{\mathrm{L}} + EA_{\mathrm{s}}} \cdot V_{\mathrm{Ed}} \\ V_{\mathrm{Ed}} - V_{\mathrm{Rds}} \end{cases} \tag{3.69}$$

### 3.4.2 Shear strengthening

If the shear strength of the member to be strengthened is inadequate, the shear capacity in the DAfStb guideline [1, 2] can be increased with the help of shear strengthening. Shear strengthening includes full wrapping and U-wrapping, as shown in Figure 3.10. Shear strengthening in the form of U-wrapping is only permitted on beams with a rectangular cross-section, not on T-beams. Although a marginal increase in the shear strength of T-beams with U-wrapping has been observed in tests, a reliable mechanical model corresponding to the German level of safety is, however, not available at the moment owing to the lack of anchorage of the wrapping in the compression zone.

In principle, in the DAfStb guideline [1, 2], increasing the shear strength involves adding the component from the strengthening to the component from the shear strength of the unstrengthened member according to DIN EN 1992-1-1 [20] with its associated National Annex [21]. Both components are based on the truss model of the Eurocode with variable strut angles. The load-carrying capacity of the tie in the truss can be calculated in a simplified form in the DAfStb guideline [1, 2] using Equation 3.70. Besides the load-carrying capacity of the tie, that of the strut, which is not only directly influenced by the strengthening, must be verified according to DIN EN 1992-1-1 [20] Equation 6.9 for shear strengthening using the chosen strut angle.

$$V_{\mathrm{Rd}} = V_{\mathrm{Rd,s}} + V_{\mathrm{Rd,Lw}} \tag{3.70}$$

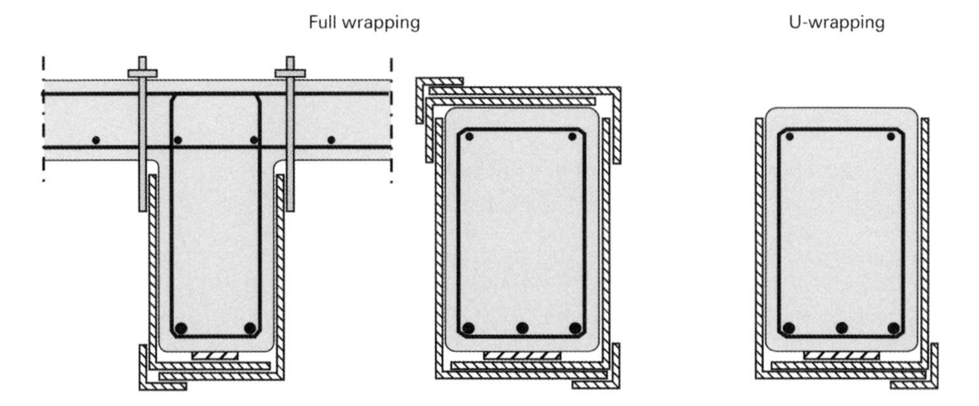


**Fig. 3.10** Potential shear strengthening schemes

Equation 3.71 is used to calculate the additional shear force that can be carried. The angle of the strut should be determined according to DIN EN 1992-1-1 [20] together with its associated National Annex [21].

$$V_{\mathrm{Rd,Lw}} = \frac{A_{\mathrm{Lw}}}{s_{\mathrm{Lw}}} \cdot z \cdot f_{\mathrm{Lwd}} \cdot \cot\theta \tag{3.71}$$

The area of the shear strengthening is calculated according to Equation 3.72 depending on the manner of applying the strengthening:

$$\frac{A_{\mathrm{Lw}}}{s_{\mathrm{Lw}}} = \begin{cases} \dfrac{2 \cdot t_{\mathrm{Lw}} b_{\mathrm{Lw}}}{s_{\mathrm{Lw}}} & \text{for strips} \\ 2 \cdot t_{\mathrm{Lw}} & \text{for full area} \end{cases} \tag{3.72}$$

The capacity of the shear strengthening $f_{\mathrm{wLd}}$ is determined depending on the material and the type of strengthening using the following equations:

- Full wrapping in steel: Equation 3.73
- Full wrapping in fibre-reinforced material: Equation 3.78
- U-wrapping in steel: Equation 3.80
- U-wrapping in fibre-reinforced material: Equation 3.80.

#### 3.4.2.1 Full wrapping in steel

The strength of full wrapping in steel is the minimum of the yield stress and the stress that can be transferred across any laps:

$$f_{\mathrm{Lwd,GS}} = \min\{f_{\mathrm{yd}}; f_{\mathrm{Gud,Lw}}\} \tag{3.73}$$

The stress that can be transferred across laps is calculated using Equations 3.74 to 3.77 depending on the thickness of the L-straps $t_{\mathrm{Lw}}$ and their modulus of elasticity $E_{\mathrm{Lw}}$ plus the length of the lap $l_{\mathrm{u,LW}}$. These equations for lap joints from the DAfStb guideline [1, 2] are based on [86].

$$f_{\mathrm{Gud,LW}} = \frac{f_{\mathrm{Guk,Lw}}}{\gamma_{\mathrm{BG}}} \tag{3.74}$$

$$f_{\mathrm{Guk,Lw}} = \begin{cases} f_{\mathrm{Guk,Lw,max}} \cdot \dfrac{l_{\mathrm{u,Lw}}}{l_{\mathrm{u,Lw,max}}} \left(2 - \dfrac{l_{\mathrm{u,Lw}}}{l_{\mathrm{u,Lw,max}}}\right) & l_{\mathrm{u,Lw}} < l_{\mathrm{u,Lw,max}} \\ f_{\mathrm{Guk,Lw,max}} & l_{\mathrm{u,Lw}} \geq l_{\mathrm{u,Lw,max}} \end{cases} \tag{3.75}$$

$$f_{\mathrm{Guk,Lw,max}} = 1.004 \cdot \sqrt{\frac{E_{\mathrm{Lw}}}{t_{\mathrm{Lw}}}} \tag{3.76}$$

$$l_{\mathrm{u,Lw,max}} = 0.121 \cdot \sqrt{E_{\mathrm{Lw}} \cdot t_{\mathrm{Lw}}} \tag{3.77}$$

#### 3.4.2.2 Full wrapping in fibre-reinforced material

The strength of full wrapping in a fibre-reinforced material is calculated using Equation 3.78:

$$f_{\mathrm{Lwd,GF}} = k_{\mathrm{R}} \cdot \alpha_{\mathrm{time}} \cdot f_{\mathrm{Ld}} \tag{3.78}$$

Where such shear wrapping is made from a CF sheet, it is the tensile strength of the fibres that governs. However, several effects mean that this strength must be reduced. On the one hand, the change in direction leads to transverse pressure on the CF sheet. On the other, the unevenness of the concrete surface and holes left behind in the concrete by dislodged aggregate lead to a loss of strength (see [11]). This is taken into account by the reduction factor $k_{\mathrm{R}}$ according to Equation 3.79:

$$k_{\mathrm{R}} = \begin{cases} 0.5 \cdot \dfrac{r_{\mathrm{c}}}{60\,\mathrm{mm}} \left(2 - \dfrac{r_{\mathrm{c}}}{60\,\mathrm{mm}}\right) & r_{\mathrm{c}} < 60\,\mathrm{mm} \\ 0.5 & r_{\mathrm{c}} \geq 60\,\mathrm{mm} \end{cases} \tag{3.79}$$

The reduction factor $k_{\mathrm{R}}$, which takes into account the reduction in the static short-term strength caused by transverse pressure, is based on an evaluation of numerous change-of-direction tests carried out on concrete and metal cross-sections [56]. The value of 0.5 for this factor represents the characteristic value of an evaluation of these tests (see [11]). As numerous different products were included in this evaluation, it could be that this value is more favourable for sheets that are not affected by transverse compression. However, a reduction to half the strength of the fabric in the case of shear strengthening is also proposed in many other international publications (see evaluations in [7] and [11]).

Moreover, the DAfStb guideline [1, 2] includes a creep rupture factor $\alpha_{\mathrm{time}}$ to take into account the effects on adhesive joints over time. This is because when using shear wrapping made from a fibre-reinforced material there are always adhesive joints or laps between various layers. According to [56] it is assumed that this factor allows for a lap length of min. 80 mm, which is guaranteed by the requirement for a 250 mm long lap in the detailing provisions of the DAfStb guideline [1, 2]. The creep rupture factor $\alpha_{\mathrm{time}}$ assumes the use of customary epoxy resins, but other values might emerge when using totally different resins.

#### 3.4.2.3 U-wrapping

When using U-wrapping, the strength of the wrapping is the minimum of the strength of full wrapping and the bond of the surface-mounted shear wrapping:

$$f_{\mathrm{Lwd}} = min\{f_{\mathrm{bLwd}}; f_{\mathrm{Lwd,G}}\} \tag{3.80}$$

The tests carried out in [11] revealed that the structural response of U-wrapping also depends very much on the position of the surface-mounted shear wrapping in relation to the internal links. If we assume that a shear crack forms at an angle of 45° and the top end of this crack is very close to one of the internal links, then an external strap attached in the vicinity of another link will have no effect, as is shown in Figure 3.11. This is because the effective length of the adhesive joint (shaded area of wrapping) becomes

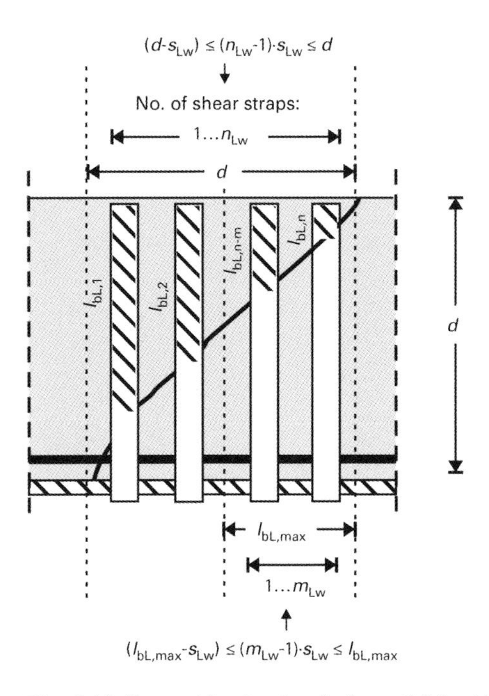


**Fig. 3.11** Geometric situation (schematic) for U-wrapping

smaller and smaller. This is taken into account by Equations 3.81 to 3.83 because as the distance between the links decreases, so the effect of an individual link is reduced further and further.

When $d \geq l_{\mathrm{bL,max}}$ and $l_{\mathrm{bL,max}} \leq s_{\mathrm{Lw}} \leq d$, then

$$f_{\mathrm{bLwd}} = \frac{f_{\mathrm{bLk,max}}}{\gamma_{\mathrm{BA}}} \tag{3.81}$$

When $d \geq l_{\mathrm{bL,max}}$ and $s_{\mathrm{Lw}} \leq l_{\mathrm{bL,max}}$, then

$$f_{\mathrm{bLwd}} = \frac{f_{\mathrm{bLk,max}}}{\gamma_{\mathrm{BA}}} \cdot \left( \left( 1 - \frac{(m_{\mathrm{Lw}} - 1)}{(n_{\mathrm{Lw}} - 1)} \right) + \frac{m_{\mathrm{Lw}} \cdot (m_{\mathrm{Lw}} - 1) \cdot s_{\mathrm{Lw}}}{2 \cdot (n_{\mathrm{Lw}} - 1) \cdot l_{\mathrm{bL,max}}} \right) \tag{3.82}$$

When $d \leq l_{\mathrm{bL,max}}$ and $s_{\mathrm{Lw}} \leq d$, then

$$f_{\mathrm{bLwd}} = \frac{f_{\mathrm{bLk,max}}}{\gamma_{\mathrm{BA}}} \cdot \frac{n_{\mathrm{Lw}} \cdot s_{\mathrm{Lw}}}{2 \cdot l_{\mathrm{bL,max}}} \tag{3.83}$$

The bond strength $f_{\mathrm{bLk,max}}$ and the effective bond length $l_{\mathrm{bL,max}}$ in the equations are determined as shown in Section 3.3.4.3.

### 3.4.3 End strap to prevent concrete cover separation failure

When using externally bonded or near-surface-mounted reinforcement, the layer of concrete directly beneath the reinforcement can become detached near the supports.

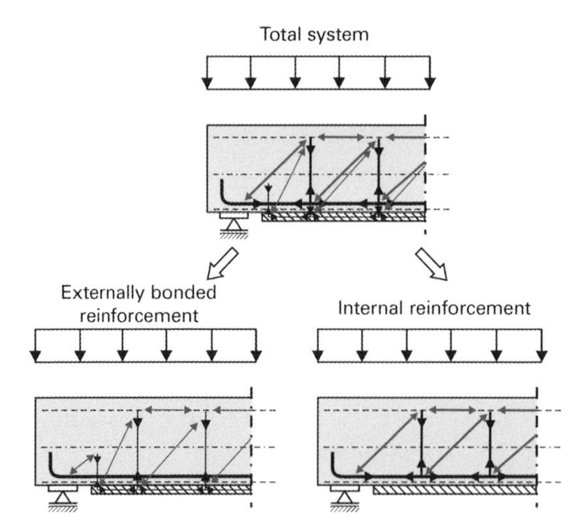


**Fig. 3.12** Cause of transverse tensile stresses at end of strip

This is known as a concrete cover separation failure. The additional, vertical offset between internal shear links and external strip leads to the build-up of tensile forces between the internal and the externally bonded reinforcement. This problem is similar to the situation with dapped supports, which are frequently encountered in precast concrete construction. The acting forces can be determined approximately with a truss model, as shown in Figure 3.12. To do this, the shear force is distributed in line with the truss according to DIN EN 1992-1-1 [20] over the strut in the concrete and the tie in the form of the internal shear link for the unstrengthened cross-section. A similar truss is now set up for the component of the tensile force in the externally bonded reinforcement and the two systems are superimposed to create one total system. A concrete cover separation failure occurs when the tie in the concrete, which results from the force in the strip, can no longer be carried by the tensile strength of the concrete.

In order to prevent this type of failure, a shear strap must be positioned at the end of the strip according to Equation 3.84 when the acting shear force at the end support, or end of the beam, is greater than the shear capacity. In Equation 3.84, $V_{\mathrm{Rd,c}}$ is determined according to DIN EN 1992-1-1 [20] in conjunction with its National Annex [21]. The variable $a_{\mathrm{L}}$ is the distance of the strip from the end support, or end of beam, in mm.

$$V_{\mathrm{Rd,c,LE}} = 0.75 \cdot \left(1 + 19.6 \cdot \frac{(100\rho_{\mathrm{s1}})^{0.15}}{a_{\mathrm{L}}^{0.36}}\right) \cdot V_{\mathrm{Rd,c}} \tag{3.84}$$

The critical value of the shear stress comes from a conversion of the model by *Jansze* [87] to the model of DIN EN 1992-1-1 [20] in conjunction with its National Annex [21]. This conversion and a validation can be found in [11].

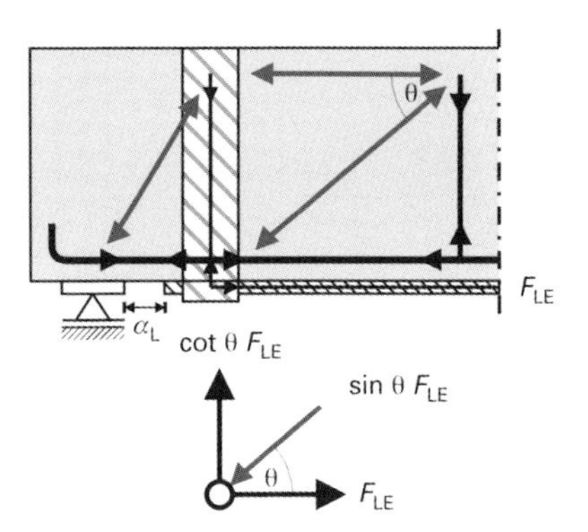


**Fig. 3.13** Calculation of force in shear wrapping due to tensile force in bending reinforcement

Where a shear strap is necessary, it should be designed according to Figure 3.13 for the design value of the acting tensile force according to Equation 3.85:

$$F_{\mathrm{LwEd}} = F^{*}_{\mathrm{LEd}} \cdot \tan\theta \tag{3.85}$$

When using externally bonded reinforcement, an approach that lies on the safe side is to design the force at the end strap for the maximum force in the flexural strengthening that can be accommodated by the end anchorage according to Equation 3.86. This is because a larger force cannot occur in the strip at the end strap except when the shear strap counts towards increasing the bond force of the flexural strengthening. The force in the shear strap is, however, superposed on the maximum force that can be accommodated at the end anchorage according to Section 3.7.2.

$$F^{*}_{\mathrm{LEd}} = f_{\mathrm{bLk,max}} \cdot b_{\mathrm{L}} \cdot t_{\mathrm{L}} \tag{3.86}$$

## 3.5 Fatigue analysis

When checking fatigue for non-static loads, the DAfStb guideline can be used to analyse the bond of the flexural strengthening in the form of externally bonded CFRP strips. As the carbon fibres exhibit virtually no signs of fatigue, only the bond needs to be checked for fatigue when using CFRP strips. Besides the fatigue of the strengthening system, the concrete, reinforcing steel and prestressing steel must also be checked according to DIN EN 1992-1-1 [20] in conjunction with its National Annex [21].

To verify the bond of externally bonded CFRP strips, the DAfStb guideline calls for a quasi-fatigue strength analysis by limiting the change in force in the strip at the concrete element between cracks. This approach was developed in DAfStb publication 593 [10] (q.v. [88]) on the basis of numerous tests [10, 89, 90]. The DAfStb guideline contains both a simplified analysis and also a more accurate approach. In the former it must be shown that owing to the bond stresses occurring as a result of the maximum load, the elastic range of the bilinear bond stress–slip model is not exceeded. In the latter, the stress ranges must be investigated, which, however, as with the fatigue of the concrete, depend on the mean stress. This means that as the minimum load rises, so the fatigue stress range diminishes.

## 3.6 Analyses for the serviceability limit state

The DAfStb guideline, supplementing DIN EN 1992-1-1 [20] together with its National Annex [21], contains expressions for limiting stresses, crack widths and deformations when assessing the serviceability limit state.

At the serviceability limit state, the strain in the strengthening system is limited for reasons of the durability of the bond. The intention behind limiting strain to 2 mm/m for rare load combinations is to prevent significant damage to the bond of the externally bonded reinforcement. Furthermore, yielding of the reinforcing steel under rare load combinations is ruled out, likewise to prevent high bond stresses and irreversible deformations of the structure.

The DAfStb guideline does not generally require verification that crack widths are not excessive. However, adhesively bonded reinforcement can limit crack widths. To exploit this effect, the guideline provides an accurate calculation method based on [91].

According to the DAfStb guideline, the limit values for member deformation to DIN EN 1992-1-1 [20] in conjunction with its National Annex [21] may not be exceeded, even after strengthening. When using the more accurate approach of DIN EN 1992-1-1 [20] section 7.4.3, the bonded reinforcement can also be included when calculating the deflection parameter. It is not possible to apply the simplified method to calculate the deflection of strengthened members according to DIN EN 1992-1-1 [20] section 7.4.2. However, aids (see [92]) may be used when working with the more accurate method.

## 3.7 Detailing

### 3.7.1 Strip spacing

When using externally bonded CFRP strips, the centre-to-centre spacing $a_L$ of the tension strips may not exceed the values according to Equation 3.87, which ensures that the strips act uniformly over the width of the member:

$$\max a_L \leq \begin{cases} 0.2 \times \text{effective span} \\ 5 \times \text{slab depth} \\ 0.4 \times \text{length of cantilever} \end{cases} \tag{3.87}$$

The values for this are included in all the former national technical approvals since the first approval [93] for externally bonded steel plates and have proved to be worthwhile ever since. Besides this requirement, it is also specified that the CFRP strip closest to the edge of a member should be positioned no closer to that edge than a distance equal to the concrete cover $c_{nom}$ to the internal reinforcement. The intention of this is to guarantee that there is no spalling along the edge and that the strip forces are transferred uniformly into the member in the region of the internal steel shear links.

In contrast to near-surface-mounted reinforcement, no minimum spacing rules apply when using externally bonded reinforcement because the bond forces are transferred directly into the surface underneath the adhesive and there is no significant spreading of the bond stress over the width of the member.

### 3.7.2 Provision of shear straps

The provision of shear straps for the adhesively bonded reinforcement can be necessary for any of the reasons explained in the preceding sections. Generally, it is necessary to distinguish between three cases for externally bonded shear straps:

***Case 1:*** The externally bonded shear strap is required for the shear design. Straps for case 1 can also serve as end straps to prevent a concrete cover separation failure. With such a strap, it is also possible to take into account an increase in the bond force due to the provision of shear wrapping. The effects due to the shear force, the provision of shear wrapping and the crack opening force $F_u$ ($\alpha_b$) should be superposed in this situation.

***Case 2:*** The externally bonded shear strap serves as an end strap or is required because of the shear wrapping according to Section 3.4.1. An increase in the bond force may be taken into account with this strap. In this situation the effects due to the provision of shear wrapping are to be superposed on the effect due to the crack opening force $F_u$ ($\alpha_b$).

***Case 3:*** The externally bonded shear strap is provided because of an increase in the bond force. This strap may be positioned anywhere and is to be designed for the crack opening force $F_u$ ($\alpha_b$) according to Section 3.3.4.4.

### 3.7.3 Steel shear straps

Shear straps made from steel plates are normally made up of two L-plates to make them easier to install. To ensure that deformation of the member does not lead to large forces/stress normal to the adhesive surface at the lap joint, [86] requires that a closing L-section (see Figure 3.14) be included to prevent the plates from coming apart. When using the shear strap to increase the bond force at the end anchorage of flexural strengthening according to Section 3.3.4.4, the DAfStb guideline specifies additional geometrical requirements to be met by the individual L-plates due to the definition of the values $l_1$ to $l_4$ according to [82].

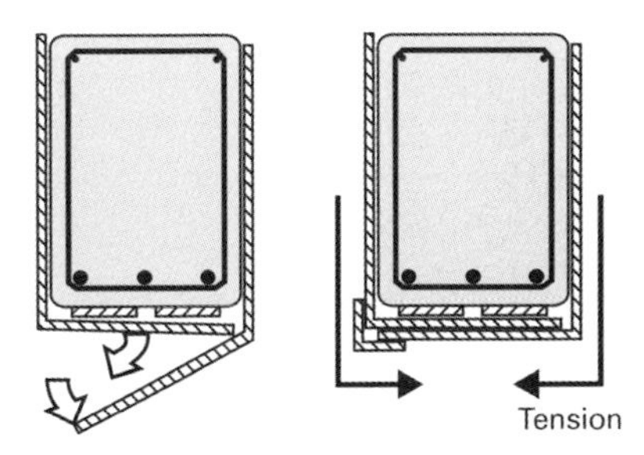


**Fig. 3.14** Function of the closing L-section

# 4 Example 1: Strengthening a slab with externally bonded CFRP strips

## 4.1 System

### 4.1.1 General

The example presented on the following pages is dealt with in considerable detail to aid understanding, and in some cases several analysis options are described. It has therefore something of a textbook character and does not represent the approach that would be chosen in practice for structural calculations.

Owing to a change of use, a reinforced concrete floor slab in a residential building needs to be strengthened. The structure was built in the year 2000 and as-built documents are available. Externally bonded CFRP strips are to be used as the strengthening system. The floor slab spans one way and was designed as a simply supported member. It is assumed that the slab is free to rotate at its supports on the masonry walls. The slab is not designed to act as a horizontal diaphragm for stability purposes. Dry internal conditions are assumed. Figure 4.1 shows the structural system requiring strengthening.

### 4.1.2 Loading

The loads are predominantly static. Three load cases will be investigated for ultimate limit state design:

- **Load case 1** represents the situation prior to strengthening.
- **Load case 2** is the loading during strengthening. The strengthening measures are carried out under the dead load of the slab. Existing fitting-out items will be removed during the strengthening work.
- **Load case 3** represents the loading situation in the strengthened condition.

Table 4.1 lists the actions of the various load cases for the loads given in Figure 4.1.

Load case 3 governs for designing the strengthening measures. The load combination for the ultimate limit state and the load combination for the serviceability limit state under a rare load combination are required for the analyses. These load combinations are in line with the requirements of DIN EN 1990 [24] together with its associated National Annex [25]. The following applies for the ultimate limit state (persistent and transient design situations):

$$\sum_{j\geq 1} \gamma_{G,j} \cdot G_{k,j} + \gamma_P \cdot P + \gamma_{Q,1} \cdot Q_{k,1} + \sum_{i>1} \gamma_{Q,i} \cdot \psi_{0,i} \cdot Q_{k,i}$$

$$p_d = \gamma_G \cdot \left(g_{1,k} + g_{2,k}\right) + \gamma_Q \cdot q_k = 1.35 \cdot (4 + 3) + 1.5 \cdot 5 = 16.95\ \mathrm{kN/m^2}$$

The load for the serviceability limit state is calculated as follows for a rare load combination:

$$\sum_{j\geq 1} G_{k,j} + P + Q_{k,1} + \sum_{i>1} \psi_{0,i} \cdot Q_{k,i}$$

*Strengthening of Concrete Structures with Adhesively Bonded Reinforcement: Design and Dimensioning of CFRP Laminates and Steel Plates*. First Edition. Konrad Zilch, Roland Niedermeier, and Wolfgang Finckh.

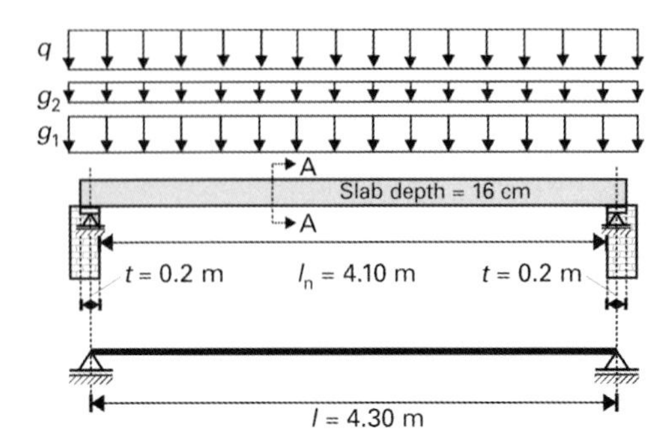


**Fig. 4.1** System for strengthening the slab in the example

**Table 4.1** Loads on the system in kN/m² for the various load cases.

| Load case | 1 | 2 | 3 |
|---|---|---|---|
| $g_{1,k}$ (dead load) | 4.0 | 4.0 | 4.0 |
| $g_{2,k}$ (fitting-out load) | 2.0 | — | 3.0 |
| $q_k$ (imposed load, category A) | 2.0 | — | 5.0 |

$$p_{rare} = g_{1,k} + g_{2,k} + q_k = 4 + 3 + 5 = 12\ \mathrm{kN/m^2}$$

In order to determine the prestrain condition during strengthening, which according to DAfStb guideline [1, 2] part 1 section 5.1.1 (RV 19) must be considered for a quasi-permanent load combination, we get the following for load case 2:

$$\sum_{j\geq 1} G_{k,j} + P + \sum_{i\geq 1} \psi_{2,i} \cdot Q_{k,i}$$

$$p_{perm} = g_{1,k} = 4\ \mathrm{kN/m^2}$$

### 4.1.3 Construction materials

#### 4.1.3.1 Near-surface tensile strength

DAfStb guideline [1, 2] part 1 section 3.1.2 (RV 10) requires the near-surface tensile strength of the member to be determined. Five values are found by testing, which are given in Table 4.2.

According to DAfStb guideline [1, 2] part 4 annex A, the value expected for the mean of the near-surface tensile strength must be determined for the design from the random sample of five values:

$$f_{ctm,surf} = f_m = \left(\frac{1}{n} \cdot \sum_{i=1}^{n} f_i\right) - k \cdot s = 2.28 - 0.953 \cdot 0.19 = 2.1\ \mathrm{N/mm^2}$$

**Table 4.2** Individual values obtained *in situ* for near-surface tensile strength.

| $n$ | 1 | 2 | 3 | 4 | 5 |
|---|---|---|---|---|---|
| $f_{ctm,surf,i}$ [N/mm$^2$] | 2.2 | 2.5 | 2.0 | 2.4 | 2.3 |
| Mean value: 2.28 N/mm$^2$ | | | | | |
| Standard deviation: 0.19 | | | | | |

#### 4.1.3.2 Concrete compressive strength

Concrete of class B25 was able to be ascertained from the as-built documents according to DIN 1045 [94]. Following a test on the member according to DIN EN 13 791 [95], the result was a strength class of C20/25. Therefore, the values according to DIN EN 1992-1-1 [20] Tab. 3.1 for C20/25 concrete will be used for the design. This results in a mean concrete compressive cylinder strength $f_{cm} = 28$ N/mm$^2$ and a characteristic concrete compressive cylinder strength $f_{ck} = 20$ N/mm$^2$.

#### 4.1.3.3 Type and quantity of existing reinforcement

The as-built documentation reveals that a type R 443 steel mesh was used as the reinforcement. Therefore, according to [96], this is a type BSt 500 M (IV M) steel reinforcing mesh to DIN 1045 [94] or DIN 488-2 [97]. Consequently, we can assume a yield stress $f_{yk} = 500$ N/mm$^2$ and a modulus of elasticity $E_s = 200$ kN/mm$^2$. As given in [96], a type R 443 mesh has longitudinal bars with an area $a_{sl} = 4.43$ cm$^2$/m, which consists of pairs of Ø6.5 mm bars @ 150 mm c/c, and transverse bars with $a_{sq} = 0.95$ cm$^2$/m, made up of Ø5.5 mm bars @ 250 mm c/c.

#### 4.1.3.4 Position of existing reinforcement

The as-built documents indicate a concrete cover of min $c = 1.0$ cm, or nom $c = 2.0$ cm, according to DIN 1045 [94]. A survey according to [98] has revealed that the reinforcement is positioned as shown in Figure 4.2.

#### 4.1.3.5 Strengthening system

Commercially available externally bonded CFRP strips with a characteristic tensile strength $f_{Luk} = 2200$ N/mm$^2$ and modulus of elasticity $E_L = 170$ kN/mm$^2$ are to be used for the strengthening. According to the manufacturer, strips with dimensions of ($t_L \times b_L$) 50 × 1.4 mm, 80 × 1.4 mm and 100 × 1.4 mm are currently available ex stock. In order

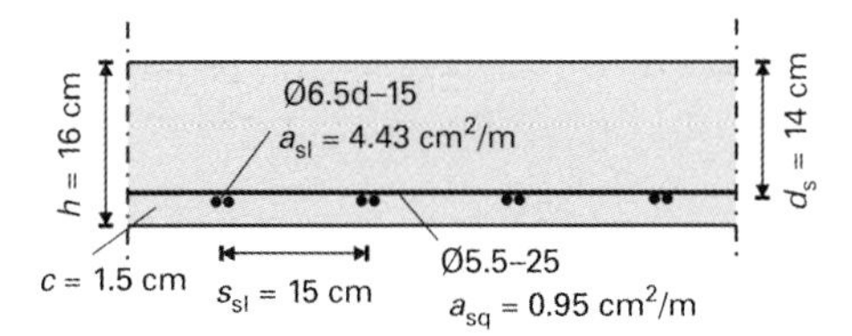


**Fig. 4.2** Type and position of existing reinforcement

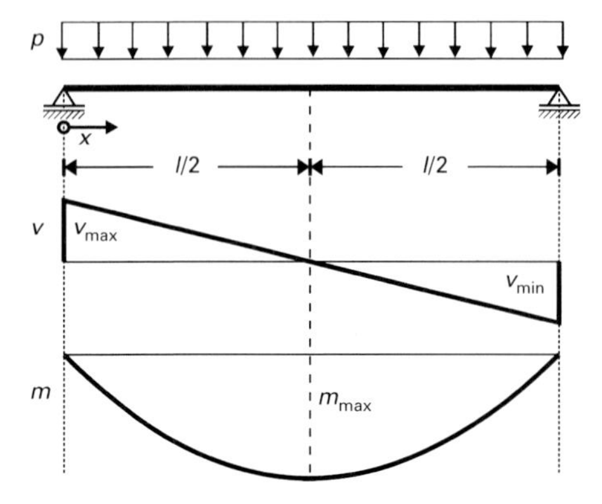


**Fig. 4.3** Shear forces and bending moments

that the work can be carried out without delay, the design will be based on these sizes. The strengthening system includes an appropriate adhesive based on epoxy resin.

## 4.2 Internal forces

Figure 4.3 shows the basic bending moment and shear force diagrams for the simply supported slab. The actual maximum values for the load combinations relevant to the design are given in Table 4.3.

$$m(x) = \frac{p}{2} \cdot l \cdot x - \frac{p \cdot x^2}{2}$$

$$v(x) = \frac{p}{2} \cdot l - p \cdot x$$

## 4.3 Determining the prestrain

DAfStb guideline [1, 2] part 1 section 5.1.1 (RV 19) requires that the prestrain be taken into account in the design. This is determined below using the example of the maximum moment. As according to the DAfStb guideline a prestrain should be determined with a quasi-permanent load combination for the serviceability limit state, characteristic material parameters are used in this section.

**Table 4.3** Maximum shear forces and bending moments for the relevant load combinations.

| Load combination | $m_{max}$ | $v_{max}$ | $v_{min}$ |
|---|---|---|---|
| — | kNm/m | kN/m | kN/m |
| Load case 3; ULS | 39.18 | 36.44 | −36.44 |
| Load case 3; SLS, rare | 27.74 | 25.80 | −25.80 |
| Load case 2; SLS, quasi-permanent | 9.25 | 8.60 | −8.60 |

An iterative method is used to determine the prestrain condition in the cross-section. The calculation below uses the internal lever arm of the reinforcing steel, determined iteratively, in order to demonstrate the method briefly. The internal lever arm, which represents the iteration variable, is

$$z_{s1} \approx 0.926 \cdot d_{s1} \approx 0.926 \cdot 140 \approx 129.60 \text{ mm}$$

The tensile force in the steel at the time of strengthening for the maximum moment can be calculated from the moment and the internal lever arm (see Section 3.2 and Figure 3.3):

$$F_{s1} = \frac{m_{0,k}}{z_{s1}} = \frac{9.25 \cdot 10^6}{129.60} = 71.34 \text{ kN/m}$$

Following on from that it is possible to determine the strain in the reinforcing steel from the area of the reinforcing bars and the modulus of elasticity of the reinforcement:

$$\varepsilon_{s1} = \frac{F_{s1}}{a_{s1} \cdot E_s} = \frac{71.34 \cdot 10^3}{4.43 \cdot 10^2 \cdot 200} = 0.75 \text{ mm/m}$$

Assuming a compressive strain in the concrete $\varepsilon_c > -2$ mm/m, the compressive force in the concrete according to Section 3.2 can be calculated approximately using the parabola-rectangle diagram for concrete under compression as follows:

$$F_c = b \cdot x \cdot f_{ck} \cdot \alpha_R = b \cdot \xi \cdot d_{s1} \cdot f_{ck} \cdot \left(-\frac{\varepsilon_c^2}{12} - \frac{\varepsilon_c}{2}\right)$$

$$= 1000 \cdot \left(\frac{-\varepsilon_c}{-\varepsilon_c + \varepsilon_{s1}}\right) \cdot 140 \cdot 20 \cdot \left(-\frac{\varepsilon_c^2}{12} - \frac{\varepsilon_c}{2}\right)$$

Equilibrium of the internal forces results in an equation for calculating the compressive strain in the concrete:

$$F_{s1} = F_c$$

$$71.34 \text{ kN/m} = -1000 \cdot \left(\frac{-\varepsilon_c}{-\varepsilon_c + 0.75}\right) \cdot 140 \cdot 20 \cdot \left(-\frac{\varepsilon_c^2}{12} - \frac{\varepsilon_c}{2}\right)$$

Solving the equation in the permissible range of values results in $\epsilon_c = -0.21$ mm/m. As this value is $> -2$ mm/m, the above assumption was justified. The relative depth of the compression zone can now be determined with the help of the strains:

$$\xi = \frac{-\varepsilon_c}{-\varepsilon_c + \varepsilon_s} = \frac{0.21}{0.21 + 0.75} = 0.22$$

Using the coefficient $k_a$ (for $\varepsilon_c > -2$ mm/m), i.e. the result according to Section 3.2, it is now possible to determine the internal lever arm:

$$k_a = \frac{8 + \varepsilon_c}{24 + 4 \cdot \varepsilon_c} = \frac{8 - 0.21}{24 - 4 \cdot 0.21} = 0.34$$

$$a = k_a \cdot \xi \cdot d_{s1} = 0.34 \cdot 0.22 \cdot 140 = 10.41 \text{ mm}$$

$$z_{s1} = d_{s1} - a = 140 - 10.41 = 129.59 \text{ mm}$$

As the internal lever arm is almost identical with the assumed lever arm, the moment of resistance for the reinforced concrete cross-section is the same as the acting moment:

$$m_{\mathrm{Rd},0} = z_{s1} \cdot F_{s1} = 129.59 \cdot 71.34 \cdot 10^{-3} = 9.25\,\mathrm{kNm/m}$$

The prestrain for the concrete therefore amounts to $\varepsilon_{c,0} = -0.21$ mm/m, and for the reinforcing steel $\varepsilon_{s1,0} = 0.75$ mm/m.

## 4.4 Simplified analysis

According to DAfStb guideline part 1, RV 6.1.1.2, and Section 3.3.2 of this book, we can carry out a simplified bond analysis. Compliance with the following boundary conditions is necessary for this analysis:

- No prestressed concrete
- Member reinforced with ribbed bars
- Strengthening in the span
- Longitudinal reinforcement not curtailed.

All these boundary conditions are satisfied in this example. The following condition must also be complied with:

$$f_{\mathrm{ctm,surf}} \geq 0.26 \cdot f_{\mathrm{cm}}^{2/3}$$

As the near-surface tensile strength is lower than this, the compressive strength of the concrete is adjusted to $f_{\mathrm{cm}} = 22.95\,\mathrm{N/mm^2}$ for determining the ultimate strain. The ultimate strain is therefore

$$\begin{aligned}
\varepsilon_{\mathrm{Ld,max}} &= 0.5\,\mathrm{mm/m} + 0.1\,\mathrm{mm/m} \cdot \frac{l_0}{h} - 0.04\,\mathrm{mm/m} \cdot \phi_s + 0.06\,\mathrm{mm/m} \cdot f_{\mathrm{cm}} \\
&= 0.5\,\mathrm{mm/m} + 0.1\,\mathrm{mm/m} \cdot \frac{4300}{160} - 0.04\,\mathrm{mm/m} \cdot 6.5 \cdot \sqrt{2} \\
&\quad +0.06\,\mathrm{mm/m} \cdot 22.95 \\
\varepsilon_{\mathrm{Ld,max}} &= 4.20\,\mathrm{mm/m}
\end{aligned}$$

In order to avoid having to perform an additional analysis for the flexural strength, a check is carried out to establish whether the strip strength is exceeded by complying with the ultimate strain. As can be seen from the following equation, this is not the case in this example and the simplified analysis is sufficient on its own.

$$\varepsilon_{\mathrm{Lud}} = \frac{f_{\mathrm{Luk}}}{E_{\mathrm{L}} \cdot \gamma_{\mathrm{LL}}} = \frac{2200}{170 \cdot 1.2} = 10.78\,\mathrm{mm/m} \geq \varepsilon_{\mathrm{Ld,max}} = 4.20\,\mathrm{mm/m}$$

The cross-sectional area of CFRP strip required was estimated beforehand iteratively using the analysis carried out here. This resulted in a strip area of approx. $a_{\mathrm{L}} = 160\,\mathrm{mm^2}$ for a strip thickness $t_{\mathrm{L}} = 1.4$ mm. As according to DAfStb guideline part 1, RV 8.2.1.1 (RV 1), or Section 3.7.1 of this book, the centre-to-centre spacing may not exceed five times the slab depth and only certain strip sizes are available, the result is the following strip area:

$$a_L = \frac{t_L \cdot b_L}{5 \cdot h} \cdot 1\,\text{m} = \frac{1.4 \cdot 100}{5 \cdot 160} \cdot 1000 = 175\,\text{mm}^2/\text{m}$$

In the rest of this example it is assumed that the strain in the strip is reached without the concrete compression zone in the cross-section failing. Owing to the continuous longitudinal reinforcement and the distribution of the bending moment, the analysis is only carried out at mid-span for the maximum moment.

The admissible tensile force in the CFRP strip is therefore given by the ultimate strain, the modulus of elasticity and the strip area:

$$F_{\text{Ld}} = a_L \cdot \varepsilon_{\text{Ld,max}} \cdot E_L = 4.20 \cdot 175 \cdot 170 = 124.86\,\text{kN/m}$$

The prestrain condition at the level of the strip is given by the prestrain determined in Section 4.3:

$$\varepsilon_{L,0} = \varepsilon_{s1,0} + \frac{d_L - d_{s1}}{d_{s1}} \cdot \left(\varepsilon_{s1,0} - \varepsilon_{c,0}\right) = 0.75 + \frac{160 - 140}{140} \cdot (0.75 + 0.21) = 0.88\,\text{mm/m}$$

The total strain at the bottom edge of the cross-section is therefore

$$\varepsilon_{L,0} + \varepsilon_{\text{Ld,max}} = 0.88 + 4.20 = 5.08\,\text{mm/m}$$

As a result of this strain, which is twice the yield strain of grade BSt 500 steel, it is assumed that the reinforcing steel is yielding. The tensile force in the reinforcing steel is therefore

$$F_{\text{s1d}} = \frac{a_{s1} \cdot f_{yk}}{\gamma_s} = \frac{4.43 \cdot 10^2 \cdot 500}{1.15} = 192.61\,\text{kN/m}$$

Assuming a compressive strain in the concrete $\varepsilon_c > -2$ mm/m leads to the following expression for the compressive force in the concrete according to Section 3.2:

$$F_c = b \cdot x \cdot f_{cd} \cdot \alpha_R = b \cdot \xi \cdot d_L \cdot \alpha_{cc} \cdot \frac{f_{ck}}{\gamma_c} \cdot \left(-\frac{\varepsilon_c^2}{12} - \frac{\varepsilon_c}{2}\right)$$

$$= 1000 \cdot \left(\frac{-\varepsilon_c}{-\varepsilon_c + \varepsilon_{L,0} + \varepsilon_{\text{Ld,max}}}\right) \cdot 160 \cdot 0.85 \cdot \frac{20}{1.5} \cdot \left(-\frac{\varepsilon_c^2}{12} - \frac{\varepsilon_c}{2}\right)$$

Equilibrium of the internal forces results in an equation to calculate the compressive strain in the concrete:

$$F_{\text{s1d}} + F_{\text{Ld}} = F_{\text{cd}}$$

Solving the equation results in $\varepsilon_c = -1.84$ mm/m. As this value is $> -2$ mm/m, the above assumption was justified. The relative depth of the compression zone can now be determined with the help of the strains:

$$\xi = \frac{-\varepsilon_c}{-\varepsilon_c + \varepsilon_{L,0} + \varepsilon_{\text{Ld,max}}} = \frac{1.84}{1.84 + 5.08} = 0.27$$

Using the coefficient $k_a$ (for $\varepsilon_c > -2$ mm/m), which is the result according to Section 3.2, it is now possible to determine the internal lever arms:

$$k_a = \frac{8 + \varepsilon_c}{24 + 4 \cdot \varepsilon_c} = \frac{8 - 1.84}{24 - 4 \cdot 18.4} = 0.37$$

$$a = k_a \cdot \xi \cdot d_L = 0.37 \cdot 0.27 \cdot 160 = 16.06 \text{ mm}$$

$$z_{s1} = d_{s1} - a = 140 - 16.06 = 123.94 \text{ mm}$$

$$z_L = h - a = 160 - 16.06 = 143.94 \text{ mm}$$

The moment capacity of the strengthened reinforced concrete cross-section is therefore

$$m_{Rd} = z_{s1} \cdot F_{s1d} + z_L \cdot F_{Ld} = (123.94 \cdot 192.61 \cdot 10^{-3} + 172.2 \cdot 143.94 \cdot 10^{-3})$$
$$= 41.85 \text{ kNm/m}$$

As the moment capacity is greater than the acting moment of 39.18 kNm/m, the design is verified.

## 4.5 Accurate analysis

### 4.5.1 General

The CFRP strip cross-section required was estimated iteratively via the bond analysis for the concrete element between cracks. That resulted in a strip thickness $t_L = 1.4$ mm with a strip area of approx. $a_L = 130 \text{ mm}^2$. As according to DAfStb guideline part 1, RV 8.2.1.1 (RV 1), or Section 3.7.1 of this book, the centre-to-centre spacing may not exceed five times the slab depth and only certain strip sizes are available, the following strip area will be chosen:

$$a_L = \frac{t_L \cdot b_{L,\text{single strip}}}{5 \cdot h} \cdot 1 \text{ m} = \frac{1.4 \cdot 80}{5 \cdot 160} \cdot 1000 = 140 \text{ mm}^2/\text{m}$$

$$b_L = \frac{a_L}{t_L} = 100 \text{ mm/m}$$

### 4.5.2 Verification of flexural strength

As with the simplified analysis, this analysis is only carried out for the maximum moment at mid-span. In the following calculation it is assumed that the reinforcing steel is yielding and the compression zone of the cross-section is fully utilized. Therefore, as with the simplified analysis, the admissible tensile force in the reinforcing steel is

$$F_{s1d} = \frac{a_{s1} \cdot f_{yk}}{\gamma_s} = \frac{4.43 \cdot 10^2 \cdot 500}{1.15} = 192.61 \text{ kN/m}$$

The strain in the strip is unknown and so the tensile force in the CFRP strip must be described in relation to this:

$$F_{\mathrm{Ld}} = a_{\mathrm{L}} \cdot \varepsilon_{\mathrm{L}} \cdot E_{\mathrm{L}} = \varepsilon_{\mathrm{L}} \cdot 140 \cdot 170\,000$$

The prestrain condition at the level of the strip is $\varepsilon_{\mathrm{L},0} = 0.88$ mm/m, as with the simplified analysis. The compressive force in the concrete can be calculated as follows by assuming a compressive strain in the concrete $\varepsilon_{\mathrm{c}} = -3.5$ mm/m according to Section 3.2:

$$F_{\mathrm{cd}} = b \cdot x \cdot f_{\mathrm{cd}} \cdot \alpha_{\mathrm{R}} = b \cdot \xi \cdot d_{\mathrm{L}} \cdot f_{\mathrm{ck}} \cdot \frac{\alpha_{\mathrm{cc}}}{\gamma_{\mathrm{c}}} \cdot \left(1 + \frac{2}{3 \cdot \varepsilon_{\mathrm{c}}}\right) =$$

$$= 1000 \cdot \left(\frac{-\varepsilon_{\mathrm{c}}}{-\varepsilon_{\mathrm{c}} + \varepsilon_{\mathrm{L}} + \varepsilon_{\mathrm{L},0}}\right) \cdot 160 \cdot 20 \cdot \frac{0.85}{1.5} \cdot \left(1 + \frac{2}{3 \cdot \varepsilon_{\mathrm{c}}}\right)$$

Equilibrium of the internal forces results in an equation to calculate the strain in the strip:

$$F_{\mathrm{s1d}} + F_{\mathrm{Ld}} = F_{\mathrm{cd}}$$

$$192.61 \cdot 10^3 + \varepsilon_{\mathrm{L}} \cdot 87.5 \cdot 170\,000$$

$$= 1000 \cdot \left(\frac{3.5}{3.5 + 0.88 + \varepsilon_{\mathrm{L}}}\right) \cdot 160 \cdot 20 \cdot \frac{0.85}{1.5} \cdot \left(1 - \frac{2}{3 \cdot 3.5}\right)$$

Iteration results in $\varepsilon_{\mathrm{L}} = 8.47$ mm/m. As this value is less than the ultimate strain in the strip (i.e. $\varepsilon_{\mathrm{Lud}} = 10.78$ mm/m), the above assumption was justified. The relative depth of the compression zone can now be determined with the help of the strains:

$$\xi = \frac{-\varepsilon_{\mathrm{c}}}{-\varepsilon_{\mathrm{c}} + \varepsilon_{\mathrm{L},0} + \varepsilon_{\mathrm{L}}} = \frac{3.5}{3.5 + 0.88 + 8.47} = 0.27$$

We can use the coefficient $k_{\mathrm{a}}$ (for $\varepsilon_{\mathrm{c}} < -2$ mm/m), which is the result according to Section 3.2, to determine the internal lever arms:

$$k_{\mathrm{a}} = \frac{3 \cdot \varepsilon_{\mathrm{c}}^2 + 4 \cdot \varepsilon_{\mathrm{c}} + 2}{6 \cdot \varepsilon_{\mathrm{c}}^2 + 4 \cdot \varepsilon_{\mathrm{c}}} = \frac{3 \cdot 3.5^2 - 4 \cdot 3.5 + 2}{6 \cdot 3.5^2 - 4 \cdot 3.5} = 0.42$$

$$a = k_{\mathrm{a}} \cdot \xi \cdot d_{\mathrm{L}} = 0.42 \cdot 0.27 \cdot 160 = 18.11 \text{ mm}$$

$$z_{\mathrm{s1}} = d_{\mathrm{s1}} - a = 140 - 18.11 = 121.89 \text{ mm}$$

$$z_{\mathrm{L}} = h - a = 160 - 18.11 = 141.89 \text{ mm}$$

The moment capacity of the strengthened reinforced concrete cross-section is therefore

$$m_{\mathrm{Rd}} = z_{\mathrm{s1}} \cdot F_{\mathrm{s1d}} + z_{\mathrm{L}} \cdot a_{\mathrm{L}} \cdot \varepsilon_{\mathrm{L}} \cdot E_{\mathrm{L}} = (121.89 \cdot 192.61 \cdot 10^{-3}$$

$$+141.89 \cdot 8.47 \cdot 140 \cdot 170 \cdot 10^{-6}) = 52.09 \text{ kNm/m}$$

As the moment capacity is greater than the acting moment of 39.18 kNm/m, the flexural strength is verified.

### 4.5.3 Determining the crack spacing

The crack spacing is required for the accurate bond analysis. This will be determined here according to DAfStb guideline part 1, RV 6.1.1.3. To do this, it is first necessary to determine the cracking moment of the cross-section with the help of the near-surface tensile strength:

$$m_{cr} = \kappa_{fl} \cdot f_{ctm,surf} \cdot W_{c,0}$$

$$\kappa_{fl} = 1.6 - \frac{h}{1000} = 1.6 - \frac{160}{1000} = 1.44 \geq 1.0$$

$$W_{c,0} = \frac{b \cdot h^2}{6} = \frac{1000 \cdot 160^2}{6} = 4.27 \cdot 10^6 \text{ mm}^3/\text{m}$$

$$m_{cr} = 1.44 \cdot 2.1 \cdot 4.27 = 12.91 \text{ kNm/m}$$

We also require the mean bond stress in the reinforcing steel in order to calculate the bond force per unit length. In doing so, the double bars of the R 443 mesh are multiplied by the factor $\sqrt{2}$ according to the DAfStb guideline:

$$f_{bsm} = \kappa_{vb1} \cdot 0.43 \cdot f_{cm}^{2/3} = 1.0 \cdot 0.43 \cdot 28^{2/3} = 3.96 \text{ N/mm}^2$$

$$F_{bsm} = \sum_{i=1}^{n} n_{s,i} \cdot \phi_{s,i} \cdot \pi \cdot f_{bsm} = \frac{443}{6.5^2 \cdot \pi/4 \cdot 2} \cdot \sqrt{2} \cdot 6.5 \cdot \pi \cdot 3.96 = 764.32 \text{ N/m}$$

Consequently, the cracking moment and the internal lever arm, which may be assumed to be 0.9 times the static effective depth, can be used to calculate the transmission length of the reinforcing steel:

$$l_{e,0} = \frac{m_{cr}}{z_s \cdot F_{bsm}} = \frac{m_{cr}}{0.9 \cdot d_{s1} \cdot F_{bsm}}$$

$$l_{e,0} = \frac{12.91 \cdot 10^6}{0.9 \cdot 140 \cdot 764.32} = 133.98 \text{ mm}$$

According to the DAfStb guideline, the crack spacing is then 1.5 times the transmission length of the reinforcing steel:

$$s_r = 1.5 \cdot l_{e,0} = 1.5 \cdot 133.98 = 200.96 \text{ mm}$$

### 4.5.4 Accurate analysis of concrete element between cracks

For this analysis, DAfStb guideline part 1, RV 6.1.1.3.6, which is described in Section 3.3.3.3 of this book, requires verification of every concrete element between cracks. It must be ensured that the acting change in force in the strip is less than the admissible change in force in the strip at the concrete element between cracks.

$$\Delta F_{LEd} \leq \Delta F_{LRd}$$

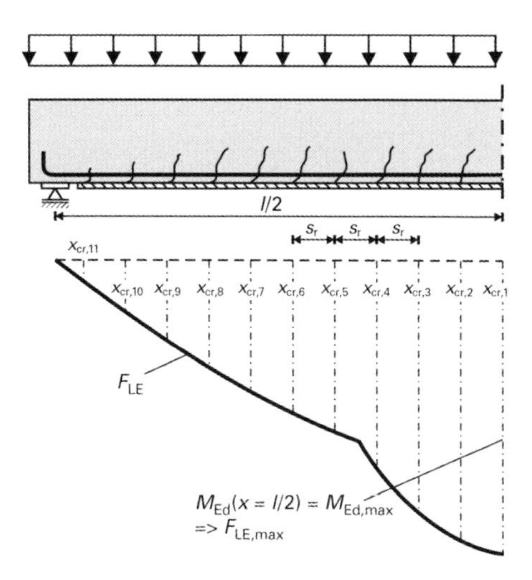


**Fig. 4.4** Concrete elements between cracks. (half span, schematic)

According to DAfStb guideline part 1, RV 6.1.1.3.6 (RV 2), the superposition principle does not apply in this analysis. However, as this example involves a statically determinate simply supported member, the load combination that produces the maximum moment is also the most unfavourable combination for checking the bond. Discrete concrete elements between cracks, starting at the maximum moment, are arranged according to the schematic drawing of Figure 4.4. First of all, the strip forces at each crack are determined and then, following calculation of the bond strength, an analysis is carried out for every concrete element between cracks.

#### 4.5.4.1 Determining the strip forces

As an example, we shall determine the stress resultants at the third crack. From Figure 4.4 the position of this crack is

$$x_{cr,3} = \frac{l}{2} - 2 \cdot s_r = \frac{4300}{2} - 2 \cdot 200.96 = 1748.08 \text{ mm} \approx 1.75 \text{ m}$$

The moment at the ultimate limit state after strengthening is required for the calculation (load case 3):

$$m_{Ed} = \frac{p \cdot l}{2} \cdot x_{cr,3} - \frac{p \cdot x_{cr,3}^2}{2} = \frac{16.95 \cdot 4.3}{2} \cdot 1.75 - \frac{16.95 \cdot 1.75^2}{2} = 37.81 \text{ kNm/m}$$

In addition, we require the moment at this point during strengthening to determine the prestrain (load case 2):

$$m_{E,0} = \frac{p \cdot l}{2} \cdot x_{cr,3} - \frac{p \cdot x_{cr,3}^2}{2} = \frac{4 \cdot 4.3}{2} \cdot 1.75 - \frac{4 \cdot 1.75^2}{2} = 8.92 \text{ kNm/m}$$

In a similar way to Section 4.3, using this moment results in a prestrain $\varepsilon_{s,0} = 0.72$ in the reinforcing steel and $\varepsilon_{c,0} = -0.20$ in the concrete. The strain at the bottom edge of the cross-section during strengthening is therefore

$$\varepsilon_{L,0} = \varepsilon_{s1,0} + \frac{d_L - d_{s1}}{d_{s1}} \cdot (\varepsilon_{s1,0} - \varepsilon_{c,0}) = 0.72 + \frac{160 - 140}{140} \cdot (0.72 + 0.20) = 0.86\,\text{mm/m}$$

The variables $\varepsilon_L = 4.04$ and $\varepsilon_c = -1.74$ were determined iteratively with the following two conditions:

$$m_{Rd} = m_{Ed}$$

$$F_{s1d} + F_{Ld} = -F_{cd}$$

The internal forces and the resistance of the cross-section are determined below in order to check these figures and to demonstrate the method of calculation. The internal compressive force in the concrete is

$$\begin{aligned} F_{cd} &= b \cdot x \cdot f_{cd} \cdot \alpha_R = b \cdot \xi \cdot d_L \cdot f_{ck} \cdot \frac{\alpha_{cc}}{\gamma_c} \cdot \left(-\frac{\varepsilon_c^2}{12} - \frac{\varepsilon_c}{2}\right) \\ &= 1000 \cdot \left(\frac{-\varepsilon_c}{-\varepsilon_c + \varepsilon_L + \varepsilon_{L,0}}\right) \cdot 160 \cdot 20 \cdot \frac{0.85}{1.5} \cdot \left(-\frac{\varepsilon_c^2}{12} - \frac{\varepsilon_c}{2}\right) \\ &= 1000 \cdot \left(\frac{1.74}{1.74 + 6.27 + 0.86}\right) \cdot 160 \cdot 20 \cdot \frac{0.85}{1.5} \cdot \left(-\frac{1.74^2}{12} + \frac{1.74}{2}\right) \\ &= 288.69\,\text{kN/m} \end{aligned}$$

The tensile forces in the strip and the reinforcing steel can be determined via the strains, modulus of elasticity and cross-sectional areas. When determining the tensile force acting on the reinforcing steel, however, it should be remembered that the reinforcement is yielding at the calculated strip strain:

$$F_{Ld} = a_L \cdot E_L \cdot \varepsilon_L = 140 \cdot 170 \cdot 4.04 = 96.08\,\text{kN/m}$$

$$F_{s1d} = \frac{a_{s1} \cdot f_{yk}}{\gamma_s} = \frac{4.43 \cdot 10^2 \cdot 500}{1.15} = 192.61\,\text{kN/m}$$

To check the iteration, the sum of the internal forces is calculated. As this equals zero, the boundary condition for the iteration is satisfied.

$$F_{s1d} + F_{Ld} - F_{cd} = 192.61 + 96.08 - 288.69 = 0$$

We can use the relative depth of the compression zone and coefficient $k_a$ (for $\varepsilon_c > -2$ mm/m), which is the result according to Section 3.2, to determine the internal lever arms:

$$\xi = \frac{-\varepsilon_c}{-\varepsilon_c + \varepsilon_{L,0} + \varepsilon_L} = \frac{1.74}{1.74 + 4.04 + 0.86} = 0.26$$

$$k_a = \frac{8 + \varepsilon_c}{24 + 4 \cdot \varepsilon_c} = \frac{8 - 1.74}{24 - 4 \cdot 17.4} = 0.37$$

**Table 4.4** Strains and internal forces at cracks.

| $x=x_{cr}$ | $m_{Ed}$ | $\varepsilon_{L0}$ | $\varepsilon_L$ | $\varepsilon_c$ | $F_{LEd}$ | $F_{sEd}$ | $F_{cEd}$ |
|---|---|---|---|---|---|---|---|
| cm | kNm/m | mm/m | mm/m | mm/m | kN/m | kN/m | kN/m |
| 215.00 | 39.18 | 0.89 | 4.44 | −1.87 | 105.57 | 192.61 | 298.18 |
| 194.90 | 38.83 | 0.88 | 4.34 | −1.84 | 103.19 | 192.61 | 295.80 |
| 174.81 | 37.81 | 0.86 | 4.04 | −1.74 | 96.08 | 192.61 | 288.69 |
| 154.71 | 36.10 | 0.82 | 3.54 | −1.58 | 84.37 | 192.61 | 276.97 |
| 134.61 | 33.70 | 0.76 | 2.87 | −1.37 | 68.25 | 192.61 | 260.85 |
| 114.52 | 30.62 | 0.69 | 2.02 | −1.12 | 48.10 | 192.61 | 240.69 |
| 94.42 | 26.85 | 0.61 | 1.72 | −0.95 | 40.92 | 169.84 | 210.74 |
| 74.33 | 22.40 | 0.51 | 1.43 | −0.77 | 33.94 | 141.25 | 175.18 |
| 54.23 | 17.27 | 0.39 | 1.09 | −0.57 | 26.00 | 108.53 | 134.51 |
| 34.13 | 11.45 | 0.26 | 0.72 | −0.37 | 17.13 | 71.72 | 88.83 |
| 14.04 | 4.95 | 0.11 | 0.31 | −0.15 | 7.35 | 30.89 | 38.22 |

**Table 4.5** Stress resultants for concrete elements between cracks.

| Element | $x_{cr,i}$ | $x_{cr,i+1}$ | $F_{LEd,2}$ | $F_{LEd,1}$ | $\Delta F_{LEd}$ |
|---|---|---|---|---|---|
| — | cm | cm | kN/m | kN/m | kN/m |
| 1 | 215.00 | 194.90 | 105.57 | 103.19 | 2.38 |
| 2 | 194.90 | 174.81 | 103.19 | 96.08 | 7.11 |
| 3 | 174.81 | 154.71 | 96.08 | 84.37 | 11.72 |
| 4 | 154.71 | 134.61 | 84.37 | 68.25 | 16.12 |
| 5 | 134.61 | 114.52 | 68.25 | 48.10 | 20.15 |
| 6 | 114.52 | 94.42 | 48.10 | 40.92 | 7.18 |
| 7 | 94.42 | 74.33 | 40.92 | 33.94 | 6.98 |
| 8 | 74.33 | 54.23 | 33.94 | 26.00 | 7.94 |
| 9 | 54.23 | 34.13 | 26.00 | 17.13 | 8.87 |
| 10 | 34.13 | 14.04 | 17.13 | 7.35 | 9.77 |
| 11 | 14.04 | 0.00 | 7.35 | 0.00 | 7.35 |

$$a = k_a \cdot \xi \cdot d_L = 0.37 \cdot 0.26 \cdot 160 = 15.70 \text{ mm}$$

$$z_{s1} = d_{s1} - a = 140 - 15.70 = 124.30 \text{ mm}$$

$$z_L = h - a = 160 - 15.70 = 144.30 \text{ mm}$$

We can therefore work out the admissible moment for the cross-section. As this corresponds to the acting moment, the second boundary condition for the iteration is also satisfied.

$$\begin{aligned} m_{Rd} &= z_{s1} \cdot F_{s1d} + z_L \cdot F_{Ld} = (124.3 \cdot 192.61 \cdot 10^{-3} + 144.3 \cdot 96.08 \cdot 10^{-3}) \\ &= 37.81 \text{ kNm/m} = m_{Ed} \end{aligned}$$

The values for the other cracks are worked out similarly (see Table 4.4).

The difference between two cracks is the stress resultant for that concrete element between cracks; the figures are given in Table 4.5. In this table, $F_{LEd,1}$ is the strip force at the less heavily stressed crack $x_1$ and $F_{LEd,2}$ that at the more highly stressed crack $x_2$.

$$\Delta F_{LEd} = F_{LEd}(x_{cr} + s_r) - F_{LEd}(x)$$

#### 4.5.4.2 Determining the bond strength

As an example, we shall only determine the resistance at element 3. The resistance at the element between cracks depends on the action and so this must be determined for every element. The element with the greatest change in force need not necessarily be the critical element. According to DAfStb guideline part 1, RV 6.1.1.3.6, or Section 3.3.3.3 of this book, the admissible change in the strip force consists of three components:

$$\Delta F_{LRd} = \frac{\Delta F_{Lk,BL} + \Delta F_{Lk,BF} + \Delta F_{Lk,KF}}{\gamma_{BA}}$$

With the strip force $F_{LEd,1}$ at the less heavily stressed crack, the first component from the basic value of the bond of externally bonded reinforcement is

$$\Delta F_{Lk,BL} = \begin{cases} \Delta F^G_{Lk,BL} - \dfrac{\Delta F^G_{Lk,BL} - \Delta F^D_{Lk,BL}}{F^D_{Lk,BL}} F_{LEd,1} & \text{for} \quad F_{LEd,1} \leq F^D_{Lk,BL} \\ \sqrt{b_L^2 \tau_{L1k} s_{L0k} E_L t_L + F^2_{LEd,1}} - F_{LEd,1} & \text{for} \quad F^D_{Lk,BL} < F_{LEd,1} \end{cases}$$

The following calculated variables are required to determine this component. First of all, the effective bond length of the externally bonded reinforcement is calculated according to DAfStb guideline part 1, RV 8.4.6 (RV 8.11), with the factor $\kappa_{Lb} = 1.128$:

$$l_{bL,max} = \frac{2}{\kappa_{Lb}} \cdot \sqrt{\frac{E_L \cdot t_L \cdot s_{L0k}}{\tau_{L1k}}}$$

Using the boundary values of the bilinear bond stress–slip relationship $s_{L0k}$, $\tau_{L1k}$ according to DAfStb guideline part 1 annex RV K 1 and the associated long-term durability coefficients $\alpha_{cc}$ and $\alpha_{ct}$ according to DIN EN 1992-1-1/NA [20] (NDP) 3.1.6 (1) and (NDP) 3.1.6 (2), the result is

$$\tau_{L1k} = 0.366 \cdot \sqrt{\alpha_{cc} \cdot f_{cm} \cdot \alpha_{ct} \cdot f_{ctm,surf}} = 0.366 \cdot \sqrt{0.85 \cdot 28 \cdot 0.85 \cdot 2.1} = 2.39\,\text{N/mm}^2$$

$$s_{L0k} = 0,20\,\text{mm}$$

$$l_{bL,max} = \frac{2}{1.128} \cdot \sqrt{\frac{170\,000 \cdot 1.4 \cdot 0.20}{2.39}} = 251.08\,\text{mm}$$

The bond strength of the externally bonded CFRP strip is also required, which is calculated according to DAfStb guideline part 1, RV 8.4.6:

$$f_{bLk,max} = \sqrt{\frac{E_L \cdot s_{L0k} \cdot \tau_{L1k}}{t_L}} = \sqrt{\frac{170\,000 \cdot 0.20 \cdot 2.39}{1.4}} = 241.30\,\text{N/mm}^2$$

Using these figures it is now possible to work out the bond resistance required at point G:

$$\Delta F^{G}_{Lk,BL} = f_{bLk}(s_r) \cdot b_L t_L$$

$$f_{bLk}(s_r) = \begin{cases} f_{bLk,max} \cdot \dfrac{s_r}{l_{bL,max}} \left(2 - \dfrac{s_r}{l_{bL,max}}\right) & s_r < l_{bL,max} \\ f_{bLk,max} & s_r \geq l_{bL,max} \end{cases}$$

$$f_{bLk}(s_r = 200.96) = 241.30 \cdot \frac{200.96}{241.30} \cdot \left(2 - \frac{200.96}{241.30}\right) = 231.68\,\text{N/mm}^2$$

$$\Delta F^{G}_{Lk,BL} = 231.68 \cdot 100 \cdot 1.4 = 32.44 \cdot 10^3\,\text{N/m} = 32.44\,\text{kN/m}$$

And the fundamental strip force with the associated bond resistance at point D is calculated likewise:

$$F^{D}_{Lk,BL} = \frac{s_{L0k} \cdot E_L \cdot b_L \cdot t_L}{s_r} - \tau_{L1k} \cdot \frac{s_r \cdot b_L}{4} = \frac{0.20 \cdot 170\,000 \cdot 100 \cdot 1.4}{200.96}$$
$$-2.33 \cdot \frac{200.96 \cdot 100}{4}$$
$$= 11.82 \cdot 10^3\,\text{N/m} = 11.82\,\text{kN/m}$$

$$\Delta F^{D}_{Lk,BL} = \sqrt{b_L^2 \cdot \tau_{L1k} \cdot s_{L0k} \cdot E_L \cdot t_L + {F^{D}_{Lk,BL}}^2} - F^{D}_{Lk,BL}$$
$$= \sqrt{100^2 \cdot 2.39 \cdot 0.20 \cdot 170\,000 \cdot 1.4 + 11820^2} - 11\,820$$
$$= 23.97 \cdot 10^3\,\text{N/m} = 23.97\,\text{kN/m}$$

As $F^{D}_{Lk,BL} = 11.82$ kN/m $< F_{LEd,1} = 84.37$ kN/m, the first component from the basic value of the bond is

$$\Delta F_{Lk,BL} = \sqrt{b_L^2 \tau_{L1k} s_{L0k} E_L t_L + F_{LEd,1}^2} - F_{LEd,1}$$

$$\Delta F_{Lk,BL} = \sqrt{100^2 \cdot 2.39 \cdot 0.20 \cdot 170\,000 \cdot 1.4 + 84\,370^2} - 84\,370$$

$$= 6.51 \cdot 10^3 \text{ N/m} = 6.51 \text{ kN/m}$$

The second component from the frictional bond is obtained with the frictional bond stress $\tau_{LFk}$ according to DAfStb guideline part 1 annex RV K 1 as follows:

$$\Delta F_{Lk,BF} = \tau_{LFk} \cdot b_L \cdot \left( s_r - \frac{2 \cdot t_L \cdot E_L}{\tau_{L1k}} \cdot \left( \sqrt{\frac{\tau_{L1k} \cdot s_{L0k}}{t_L \cdot E_L} + \frac{F_{LEd,1}^2}{b_L^2 \cdot t_L^2 \cdot E_L^2}} - \frac{F_{LEd,1}}{b_L \cdot t_L \cdot E_L} \right) \right)$$

$$\tau_{LFk} = 10.8 \cdot \alpha_{cc} \cdot f_{cm}^{-0.89} = 10.8 \cdot 0.85 \cdot 28^{-0.89} = 0.47 \text{ N/mm}^2$$

$$F_{Lk,BF} = 0,47 \cdot 100 \cdot \left( 200.96 - \frac{2 \cdot 1.4 \cdot 170\,000}{2.39} \right.$$

$$\left. \cdot \left( \sqrt{\frac{2.39 \cdot 0.20}{1.4 \cdot 170\,000} + \frac{84\,370^2}{100^2 \cdot 1.4^2 \cdot 170\,000^2}} - \frac{84\,370}{100 \cdot 1.4 \cdot 170\,000} \right) \right)$$

$$= 6,92 \cdot 10^3 \text{ N/m} = 6.92 \text{ kN/m}$$

The third component, caused by the curvature of the member, with $\kappa_k = 24.3 \cdot 10^3$ and the crack strains $\varepsilon_{Lr1}$ and $\varepsilon_{cr1}$ at the less heavily stressed crack edge, is

$$\Delta F_{Lk,KF} = s_r \cdot \kappa_k \cdot \frac{\varepsilon_{Lr1} - \varepsilon_{cr1}}{h} \cdot b_L$$

$$\Delta F_{Lk,KF} = 201.96 \cdot 24.3 \cdot 10^3 \cdot \frac{(4.89 - (-1.74)) \cdot 10^{-3}}{160} \cdot 100$$

$$= 20.24 \cdot 10^3 \text{ N/m} = 20.24 \text{ kN/m}$$

The admissible change in the strip force for element 3 is therefore

$$\Delta F_{LRd} = \frac{\Delta F_{Lk,BL} + \Delta F_{Lk,BF} + \Delta F_{Lk,KF}}{\gamma_{BA}} = \frac{6.51 + 6.92 + 20.24}{1.5} = 22.45 \text{ kN/m}$$

The analysis for element 3 is as follows:

$$\Delta F_{LEd} = 11.71 \text{ kN/m} \leq \Delta F_{LRd} = 22.45 \text{ kN/m}$$

The analysis is carried out for every concrete element between cracks. The three components of the admissible change in the strip force depend on the acting

**Table 4.6** Comparison of admissible change in strip force and change in strip force due to action.

| Element | $\Delta F_{LEd}$ | $\Delta F_{Lk,BL}$ | $\Delta F_{Lk,BF}$ | $\Delta F_{Lk,KF}$ | $\Delta F_{LRd}$ | $\Delta F_{LEd}/\Delta F_{LRd}$ |
|---|---|---|---|---|---|---|
| — | kN/m | kN/m | kN/m | kN/m | kN/m | — |
| 1 | 2.38 | 5.39 | 7.37 | 21.95 | 23.14 | 0.10 |
| 2 | 7.11 | 5.77 | 7.22 | 21.52 | 23.00 | 0.31 |
| 3 | 11.72 | 6.51 | 6.92 | 20.24 | 22.45 | 0.52 |
| 4 | 16.12 | 7.90 | 6.37 | 18.14 | 21.61 | 0.75 |
| **5** | **20.15** | **10.68** | **5.27** | **15.26** | **20.81** | **0.97** |
| 6 | 7.18 | 12.14 | 4.69 | 11.69 | 19.02 | 0.38 |
| 7 | 6.98 | 13.95 | 3.97 | 9.99 | 18.61 | 0.38 |
| 8 | 7.94 | 16.63 | 2.91 | 8.24 | 18.52 | 0.43 |
| 9 | 8.87 | 20.75 | 1.28 | 6.27 | 18.87 | 0.47 |
| 10 | 9.77 | 27.17 | 0.00 | 4.11 | 20.85 | 0.47 |
| 11 | 7.35 | 32.44 | 0.00 | 1.75 | 22.79 | 0.32 |

strip force at the less heavily stressed crack edge of the element and thus have to be recalculated for every element. Table 4.6 compares the admissible change in the strip force with the change in the strip force due to the action at every concrete element between cracks. Element 5 is the critical one; 97% of the bond capacity is utilized here.

### 4.5.5 End anchorage analysis

Verifying the end anchorage requires an analysis at the flexural crack closest to the point of contraflexure according to DAfStb guideline part 1, RV 6.1.1.4.2, which is described here in Section 3.3.4.2. It is first necessary to determine the flexural crack that is nearest the point of contraflexure, which in the case of the simply supported slab is the one closest to the support. To do this, the cracking moment of the cross-section from Section 4.5.3 is compared with the moment at the ultimate limit state after strengthening (load case 3):

$$m_{Ed}(x_{cr}) = \frac{p \cdot l}{2} \cdot x_{cr} - \frac{p \cdot x_{cr}^2}{2} = \frac{16.95 \cdot 4.3}{2} \cdot x_{cr} - \frac{16.95 \cdot x_{cr}^2}{2} = m_{cr}$$

Using this equation, the flexural crack closest to the support is located at $x_{cr} = 389.29$ mm. Consequently, the bond length of the externally bonded reinforcement can be calculated using the depth of bearing, $t = 200$ mm (see Section 4.1.1), and the distance to the edge of the support, $a_{L,t} = 50$ mm:

$$l_{bL} = x_{cr} - t/2 - a_{L,t} = 389.29 - 200/2 - 50 = 239.29 \text{ mm}$$

As in Section 4.5.4, the effective bond length and the bond strength of the externally bonded reinforcement result from the boundary values of the bilinear bond stress–slip approach $s_{L0k}$, $\tau_{L1k}$ according to DAfStb guideline part 1 annex RV K 1, which are $l_{bL,max} = 251.08$ mm and $f_{bLk,max} = 241.30$ N/mm$^2$. Using these figures it is possible to determine the bond length and associated ultimate strain in the strip required for verifying the end anchorage:

$$l_{bL,lim} = 0.86 \cdot l_{bL,max} = 0.86 \cdot 251.08 = 215.93 \text{ mm}$$

$$\varepsilon^{a}_{LRk,lim} = 0.985 \cdot \frac{f_{bLk,max}}{E_{Lm}} = 0.985 \cdot \frac{241.30}{170} = 1.40 \text{ mm/m}$$

The available bond length $l_{bL} = 239.39$ mm is greater than the bond length $l_{bL,lim} = 215.93$ mm and so the strain in the strip as well as the associated slip can be calculated using the following equations:

$$\varepsilon^{a}_{LRk}(l_{bL} = 239.29 \text{ mm}) = \varepsilon^{a}_{LRk,lim} = 1.40 \text{ mm/m}$$

$$s^{a}_{Lr}(l_{bL} = 239,29 \text{ mm}) = 0.213 \text{ mm} + \left(l_{bL} - l_{bL,lim}\right) \cdot \varepsilon^{a}_{LRk,lim}$$
$$= 0.213 \text{ mm} + (239.29 - 215.93) \cdot 1.40 = 0.246 \text{ mm}$$

First of all, the bond coefficient of the reinforcing steel must be determined in order to calculate the strain in the reinforcing steel. To do this, the variables $\kappa_{b1k} = 2.545$, $\kappa_{b2k} = 1.0$, $\kappa_{b3k} = 0.8$ and $\kappa_{b4k} = 0.2$ according to DAfStb guideline part 1 Tab. RV 6.1 are chosen for ribbed reinforcing bars and good bond conditions:

$$\kappa_{bsk} = \kappa_{b1k} \cdot \sqrt{\frac{f_{cm}^{\kappa_{b2}}}{E_s \cdot \phi^{\kappa_{b3}} \cdot (E_L \cdot t_L)^{\kappa_{b4}}}} = 2.545$$
$$\cdot \sqrt{\frac{28^{1.0}}{200\,000 \cdot \left(\sqrt{2} \cdot 6.5\right)^{0.8} \cdot (170\,000 \cdot 1.4)^{0.2}}} = 0.0036$$

The depth of the compression zone is also required. This is calculated below in simplified form according to DAfStb guideline part 1 annex L 1:

$$x = \left[-(\alpha_L \cdot \rho_L + \alpha_s \cdot \rho_{s1}) + \sqrt{(\alpha_L \cdot \rho_L + \alpha_s \cdot \rho_{s1})^2 + 2 \cdot \left(\alpha_L \cdot \rho_L \cdot \frac{d_L}{h} + \alpha_s \cdot \rho_{s1} \cdot \frac{d_{s1}}{h}\right)}\right] \cdot h$$

$$\rho_{s1} = \frac{A_{s1}}{b \cdot h} = \frac{4.43 \cdot 10^2}{1000 \cdot 160} = 0.0028$$

$$\rho_L = \frac{A_L}{b \cdot h} = \frac{140}{1000 \cdot 160} = 0.00088$$

$$\alpha_s = \frac{E_s}{E_c} = \frac{200\,000}{29\,961} = 6.68$$

$$\alpha_L = \frac{E_L}{E_c} = \frac{170\,000}{29\,961} = 5.67$$

$$x = \left[-(5.67 \cdot 0.00088 + 6.67 \cdot 0.0028) + \sqrt{(5.67 \cdot 0.00088 + 6.67 \cdot 0.0028)^2 + 2 \cdot \left(5.67 \cdot 0.00088 \cdot \frac{160}{160} + 6.67 \cdot 0.0028 \cdot \frac{140}{160}\right)}\right] \cdot 160$$

$$= 29.36\,\text{mm}$$

Using these figures it is now possible to determine the strain in the reinforcing steel:

$$\varepsilon^a_{sRk}(l_{bL}) = \kappa_{VB} \cdot \kappa_{bsk} \cdot \left(s^a_{Lr}(l_{bL})\right)^{(\alpha_N+1)/2} \cdot \left(\frac{d^a - x^a}{d^a_L - x^a}\right)^{(\alpha_N+1)/2}$$

$$\varepsilon^a_{sRk}(l_{bL}) = 1.0 \cdot 0.0036 \cdot (0.155)^{(0.25+1)/2} \cdot \left(\frac{140 - 29.36}{160 - 29.36}\right)^{(0.25+1)/2} = 1.35\,\text{mm/m}$$

To calculate the admissible moment, the internal lever arms are still required, which can be determined in simplified form via the compression zone:

$$z^a_L = h - k_a \cdot x \approx h - 0.4 \cdot x \approx 160 - 0.4 \cdot 29.36 \approx 148.26\,\text{mm}$$

$$z^a_s = d - k_a \cdot x \approx h - 0.4 \cdot x \approx 140 - 0.4 \cdot 29.36 \approx 128.26\,\text{mm}$$

Therefore, the admissible moment at the flexural crack nearest the support is

$$m_{Rd}(l_{bL}) = \varepsilon^a_{LRk}(l_{bL}) \cdot E_{Lm} \cdot A_L \cdot z^a_L \cdot \frac{1}{\gamma_{BA}} + \varepsilon^a_{sRk}(l_{bL}) \cdot E_s \cdot A_s \cdot z^a_s \cdot \frac{1}{\gamma_S}$$

$$m_{Rd}(l_{bL}) = \left(1.40 \cdot 170 \cdot 140 \cdot 148.26 \cdot \frac{1}{1.5} + 1.35 \cdot 200 \cdot 4.43 \cdot 10^2 \cdot 128.26 \cdot \frac{1}{1.15}\right) \cdot 10^{-3}$$

$$= 3.29 + 13.32 = 16.61\,\text{kNm/m}$$

The acting moment results from the position of the flexural crack closest to the support and the ‘shift rule’, which in this analysis according to DAfStb guideline part 1 section 9.3.1.1 (RV 10) may be assumed to be $h/2$ for solid slabs:

$$m_{Ed}(x_{cr}) = \frac{p \cdot l}{2} \cdot \left(x_{cr} + \frac{h}{2}\right) - \frac{p \cdot \left(x_{cr} + \frac{h}{2}\right)^2}{2} = \frac{16.95 \cdot 4.3}{2} \cdot \left(0.39 + \frac{0.16}{2}\right) - \frac{16.95 \cdot \left(0.39 + \frac{0.16}{2}\right)^2}{2} = 13.37\,\text{kNm/m}$$

As the acting moment $m_{Ed} = 13.37$ kNm/m is less than the admissible moment $m_{Rd} = 16.62$ kNm/m, the end anchorage is verified.

## 4.6 Analysis of shear capacity

The analysis of the shear capacity is carried out according to DAfStb guideline part 1 [1, 2] section 6.2.1 (RV 10) and according to DIN EN 1992-1-1 [20] with its associated National Annex [21] section 6.2.2. The design shear force is calculated according to DIN EN 1992-1-1 section 6.2.1 (8) as follows:

$$v_{Ed,red} = v_{Ed} - p_{Ed} \cdot (t/2 + d) = 36.44 - 16.95 \cdot (0.20/2 + 0.14) = 32.37\,\mathrm{kN/m}$$

The shear resistance of a member without shear reinforcement is obtained from the maximum of Eqs. (6.2a) and (6.2b) from DIN EN 1992-1-1. The shear resistance from Eq. (6.2a) is given by

$$v_{Rd,c} = \left[C_{Rd,c} \cdot k \cdot (100 \cdot \rho_l \cdot f_{ck})^{1/3} + 0.12 \cdot \sigma_{cp}\right] \cdot d$$

The following shear resistance is calculated using the variables in Eq. (6.2a) according to DIN EN 1992-1-1, or its National Annex. It should be noted here that according to DAfStb guideline part 1 section 6.2.2 (RV 7) and DIN EN 1992-1-1 Fig. 6.3, the externally bonded reinforcement may not be counted as part of the longitudinal reinforcement.

$$k = 1 + \sqrt{\frac{200}{d}} = 1 + \sqrt{\frac{200}{140}} = 2.20 \leq 2.0$$

$$\sigma_{cp} = N_{Ed}/A_c = 0$$

$$C_{Rd,c} = \frac{0.15}{\gamma_c} = \frac{0.15}{1.5} = 0.10$$

$$100 \cdot \rho_l = \frac{a_{sl}}{d} = \frac{4.43}{14} = 0.32 \leq 2\%$$

$$v_{Rd,c} = \left[0.10 \cdot 2.0 \cdot 1.0 \cdot (0.32 \cdot 20)^{1/3}\right] \cdot 140 = 51.80\,\mathrm{kN/m}$$

The minimum shear resistance of a member without shear reinforcement is given by DIN EN 1992-1-1 Eq. (6.2b) as

$$v_{Rd,c} = \left[\frac{0.0525}{\gamma_c} \cdot \sqrt{k^3 \cdot f_{ck}} + 0.12 \cdot \sigma_{cp}\right] \cdot d = \left[\frac{0.0525}{1.5} \cdot \sqrt{2^3 \cdot 20}\right] \cdot 140 = 61.98\,\mathrm{kN/m}$$

The design shear force $v_{Ed,red} = 32.37$ kN/m is less than the shear capacity $v_{Rd,c} = 61.98$ kN/m and so the shear analysis is satisfied. Checking for a concrete cover separation failure according to DAfStb guideline part 1, RV 6.2.7, described here in Section 3.4.3, is not critical in solid slabs in which the strip continues almost to the support.

## 4.7 Serviceability limit state

Analyses of crack width and deformation are not carried out in this example. It is merely verified that the necessary stresses are complied with. According to DAfStb guideline part 1 section 7.2, described in Section 3.6 of this book, the strains in the strip and the reinforcing steel must be limited as follows for a rare load combination:

$$\varepsilon_s \leq \frac{f_{yk}}{E_s} = \frac{500}{200\,000} = 2.5\ \text{mm/m}$$

$$\varepsilon_L \leq 2\ \text{mm/m}$$

Under a rare load combination, we get the following maximum moment at mid-span:

$$m_{E,rare} = \frac{p \cdot l^2}{8} = \frac{(4+3+5) \cdot 4.3^2}{8} = 27.74\ \text{kNm/m}$$

The prestrains $\varepsilon_{s1,0} = 0.75$ mm/m, $\varepsilon_{c,0} = -0.21$ mm/m and $\varepsilon_{L,0} = 0.88$ mm/m are calculated as explained in Section 4.3. The variables $\varepsilon_L = 1.41$ mm/m and $\varepsilon_c = -0.65$ mm/m are determined iteratively via the following two conditions:

$$m_R = m_{E,rare}$$

$$F_{s1} + F_L = -F_c$$

The strain in the reinforcing steel is then determined via the depth of the compression zone:

$$x = \frac{-\varepsilon_c}{-\varepsilon_c + \varepsilon_{L,0} + \varepsilon_L} \cdot d_L = \frac{0.65}{0.65 + 0.88 + 1.41} \cdot 160 = 35.46\ \text{mm}$$

$$\varepsilon_{s1} = -\varepsilon_c \cdot \frac{d - x}{x} = 0.65 \cdot \frac{140 - 35.46}{35.46} = 1.93\ \text{mm/m} \leq 2.5\ \text{mm/m}$$

As the ultimate strains for the strip and the reinforcing steel are not exceeded, the design is verified.

# 5 Design of strengthening with near-surface-mounted CFRP strips

## 5.1 Principles

Besides strengthening options with externally bonded reinforcement, the DAfStb guideline [1, 2] also includes a design concept for flexural strengthening using near-surface-mounted CFRP strips.

In this form of strengthening the CFRP strips are fitted into slots sawn or milled in the concrete and fixed with an epoxy resin adhesive. Bond tests have shown that placing the strip in a slot and the associated distributed transfer of the tensile force into the surrounding concrete results in a very favourable, robust bond behaviour. In contrast to the situation with reinforcement bonded externally to the surface, a bond failure does not take place in the layer of concrete near the surface, but rather in the high-strength adhesive. So as it is not the moderate tensile strength of the concrete that determines the loadbearing capacity, much higher bond stresses can be transferred. Apart from that, high friction stresses can be transferred once the bond strength has been exceeded.

The comparison of the theoretical characteristic bond capacity at a single crack shown in Figure 5.1, which presumes an approximately uniform effective bonded area, clearly reveals the efficient composite action of near-surface-mounted CFRP strips. In this method the tensile strength of the strips can be reached over short transmission lengths. Compared with externally bonded reinforcement, cracks in a strengthened member are not a prerequisite for generating the tensile force (see [27]).

Owing to the very effective bond behaviour of near-surface-mounted CFRP strips, which in terms of how they work is comparable with that of the reinforcing steel, the known design approaches valid for conventional reinforced concrete can be applied here with minor adjustments. The method for designing near-surface-mounted CFRP strips is attributed to *Blaschko* [27] (q.v [99].) and has been incorporated in the DAfStb guideline virtually unchanged from the earlier approvals (see [29], for example). This method only applies within the limits given in section RV 3.8 of the guideline and cannot be applied to other forms of reinforcement such as round bars.

Further background information on near-surface-mounted reinforcement can be found in [27, 100, 101].

## 5.2 Verification of flexural strength

As with externally bonded CFRP strips, the analysis of the flexural strength can be carried out in a similar way to that for a conventional reinforced concrete member by investigating the cracked cross-section. The equations given in Section 3.2 can also be used here. When using near-surface-mounted CFRP strips, it is necessary to determine the effective structural depth $d_L$ of the CFRP strips as shown in Figure 5.2.

$$d_L = h - \left(t_s - \frac{b_L}{2}\right) \tag{5.1}$$

*Strengthening of Concrete Structures with Adhesively Bonded Reinforcement: Design and Dimensioning of CFRP Laminates and Steel Plates*. First Edition. Konrad Zilch, Roland Niedermeier, and Wolfgang Finckh.

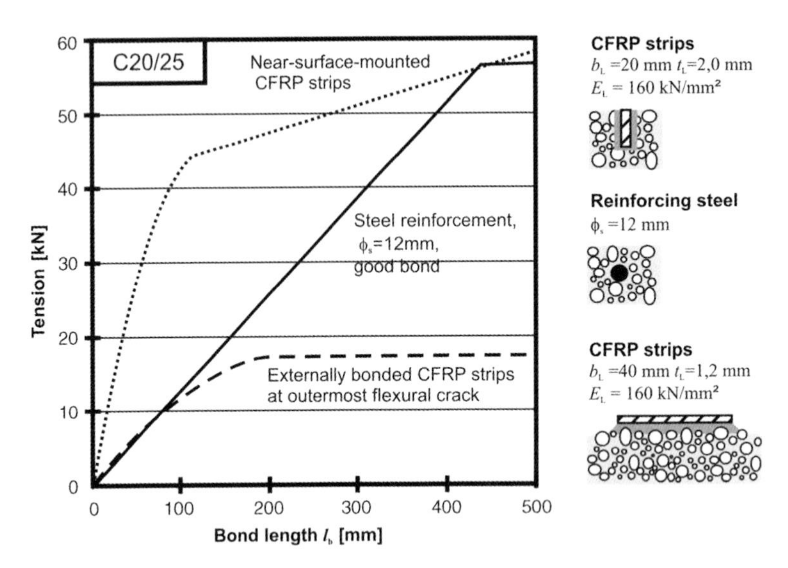


**Fig. 5.1** Bond strength of externally bonded and near-surface-mounted CFRP strips compared with ribbed reinforcing bars.(single crack)

It is assumed that the strip is always embedded at the maximum depth of the slot $t_S$ – an approach that lies on the safe side. The depth of the slot in the concrete should be dimensioned such that the strip can be fully embedded in the slot even allowing for any unevenness of the surface. According to part 3 of the DAfStb guideline, the largest permissible slot depth $t_S$ is

$$t_S \leq c - \Delta c_{\text{dev}} \tag{5.2}$$

where $\Delta c_{\text{dev}}$ is the concrete cover to the existing reinforcement. This is calculated as follows:

$$\Delta\, c_{\text{dev}} = \Delta\, c_{\text{tool}} + \Delta\, c_{\text{slot}} + \Delta\, c_{\text{member}} \tag{5.3}$$

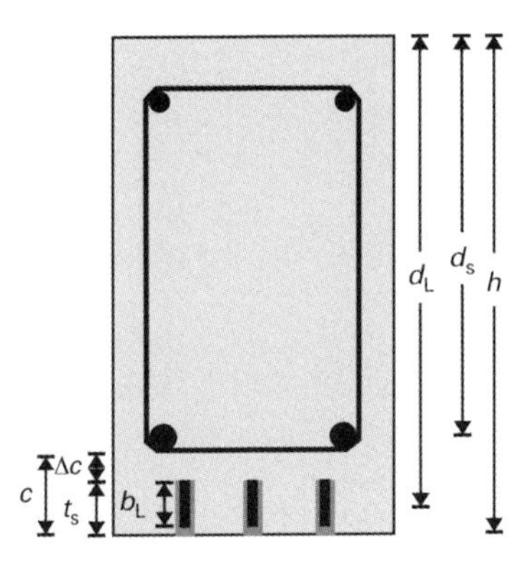


**Fig. 5.2** Cross-section with near-surface-mounted CFRP strips

where $\Delta c_{tool}$ is the tolerance specific to the tool used to measure the concrete cover according to [98] or the manufacturer's instructions and should be at least 1 mm. The allowance for the slot depth $\Delta c_{slot}$ takes into account the construction tolerances when cutting the slot; 2 mm should be selected as a minimum. In addition, $\Delta c_{member}$ is required to take into account how the concrete cover varies over the member; $\Delta c_{member} = 0$ mm may be selected for slabs, but $\Delta c_{member}$ should be at least 2 mm for all other types of member. Where the accuracy of measuring the concrete cover is improved by employing suitable measures, e.g. random checks of the cover by exposing the reinforcement locally, $\Delta c_{member}$ can be neglected.

In the DAfStb guideline the assumed ultimate strain is reduced by the factor $\kappa_{\varepsilon} = 0.8$ according to [27] when verifying the flexural strength in order to guarantee a residual minimum deformability upon reaching the theoretical bending capacity.

$$\varepsilon_{LRd,max} \leq \kappa_{\varepsilon} \cdot \varepsilon_{Lud} \tag{5.4}$$

## 5.3 Bond analysis

Essentially, the concept proposed by *Blaschko* [27] (q.v [99]) is used for verifying the bond. In this concept it is assumed that the CFRP strip makes a full contribution and there is good composite action between strip and concrete, with the strength of the adhesive usually governing this composite action. As the composite action is very effective, the full tensile strength of the CFRP strip can be anchored within a very short length – similar to conventional steel reinforcing bars. It is therefore sufficient to check the end anchorage at the point at which the strip is no longer required for the load-carrying capacity, very similar to anchoring steel reinforcing bars. This concept to describe the way in which strips in slots work has proved worthwhile over the past 10 years in the former national technical approvals and was therefore included in the DAfStb guideline.

As with conventional reinforced concrete construction with curtailed reinforcing bars, checking the bond according to the DAfStb guideline [1, 2] requires verification of the curtailment taking into account the end anchorage of the CFRP strip. This involves verifying that the design value of the member resistance is greater than the design value of the acting internal forces in the strengthened condition for every cross-section of the strengthened member. The partial tensile forces assigned to the lines of reinforcement can be determined in a simplified way by assuming a planar strain distribution. Figure 5.3 provides an overview of the curtailment verification.

As is apparent in Figure 5.3, verifying the anchorage of a near-surface-mounted CFRP strip can therefore be carried out at the point at which the CFRP strip is first required for loadbearing purposes (point A). The anchorage length $l_{bL}$ of the strip in this analysis is the distance between point A and the end of the strip. To verify the anchorage, the resulting design value of the bond capacity per CFRP strip depends on this anchorage length $l_{bL}$ and the distance of the longitudinal axis of the strip from the free edge of the member $a_r$, which may not be $>$ 150 mm, according to Equation 5.5:

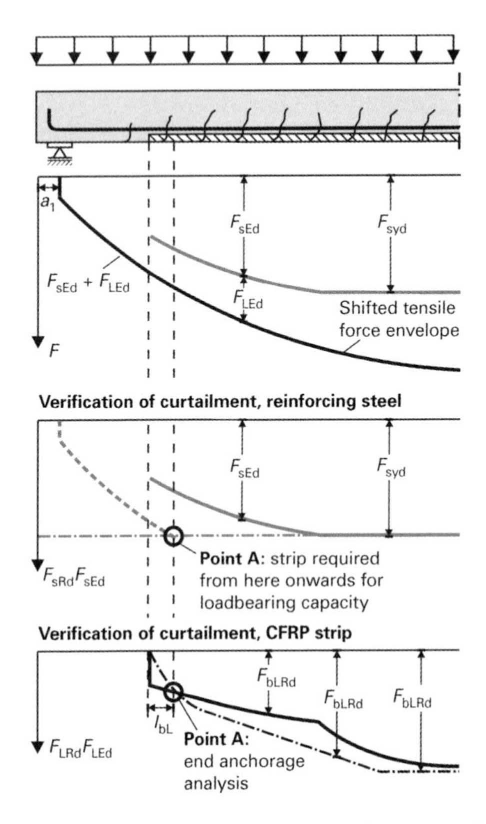


**Fig. 5.3** Verification of curtailment for near-surface-mounted CFRP strips

$$F_{bLRd} = \begin{cases} b_L \cdot \tau_{bLd} \cdot \sqrt[4]{a_r} \cdot l_{bL} \cdot (0.4 - 0.0015 \cdot l_{bL}) \cdot 0.95 & \text{for} \quad l_{bL} \leq 115\,\text{mm} \\ b_L \cdot \tau_{bLd} \cdot \sqrt[4]{a_r} \cdot \left(26.2 + 0.065 \cdot \tanh\left(\frac{a_r}{70}\right) \cdot (l_{bL} - 115)\right) \cdot 0.95 & \text{for} \quad l_{bL} > 115\,\text{mm} \end{cases} \tag{5.5}$$

To determine the bond capacity, Equation 5.5 also requires the bond strength $\tau_{bLd}$ of the near-surface-mounted CFRP strip, which according to Equation 5.6 is found from the result of the minimum of the concrete bond strength and the adhesive bond strength:

$$\tau_{bLd} = \frac{1}{\gamma_{BE}} \cdot min \begin{cases} \tau_{bGk} \cdot \alpha_{bG} \\ \tau_{bck} \cdot \alpha_{bc} \end{cases} \tag{5.6}$$

For concrete strength classes C20/25 and higher and the strengthening systems currently on the market, the bond strength of the adhesive according to Equation 5.7 governs:

$$\tau_{bGk} = k_{sys} \cdot \sqrt{\left(2 \cdot f_{Gtk} - 2 \cdot \sqrt{(f_{Gtk}^2 + f_{Gck} \cdot f_{Gtk})} + f_{Gck}\right) \cdot f_{Gtk}} \tag{5.7}$$

In this equation the bond stress is expressed in terms of the shear strength of the adhesive, which according to the Mohr-Coulomb failure criterion is given by the tensile strength $f_{Gtk}$ and the compressive strength $f_{Gck}$. However, to adjust the values from the bond tests, a system coefficient $k_{sys}$ specific to the product was incorporated in the equation. The strength of the adhesive and this system coefficient can be found in the national technical approvals for the systems and depend on the internal monitoring on the building site. If the tensile and compressive strengths are checked as part of this internal monitoring, then according to the national technical approvals for the systems, values between 21 and 28 N/mm$^2$ can be assumed for $f_{Gtk}$ and between 75 and 85 N/mm$^2$ for $f_{Gck}$. However, these characteristic values must also be obtained in the internal monitoring according to part 3 of the DAfStb guideline following a statistical evaluation. The product-specific system coefficient $k_{sys}$ lies between 0.6 and 1.0 depending on the system.

The concrete can fail in the case of a very low concrete strength and therefore the bond strength of the concrete according to Equation 5.8 governs:

$$\tau_{bck} = k_{bck} \cdot \sqrt{f_{cm}} \tag{5.8}$$

In a similar way to the bond of reinforcing steel (see [102–105], for example), this bond strength is calculated from the square root of the concrete compressive strength and a calibration factor $k_{bck}$. The system coefficient for the bond failure of the concrete can be taken from the national technical approval for the system. Tests carried out at the Technische Universität München established a characteristic value $k_{bck} = 4.5$.

The factors $\alpha_{bG}$ and $\alpha_{bC}$ were introduced into Equation 5.6 to take account of the long-term durability behaviour of the materials involved. As these are also coefficients specific to particular products, they can again be obtained from the national technical approvals. Many studies of the long-term durability behaviour of concrete have been carried out, and this behaviour is covered by DIN EN 1992-1-1 [20] together with its associated National Annex [21]. Therefore, the long-term effect coefficient $\alpha_{bC}$ for a bond failure in the concrete should lie between 0.85 and 1.0. However, adhesives can exhibit a much lower long term strength in some cases (see [100, 106–108], for example). Depending on the adhesive and the ambient conditions of the application, the long-term effect coefficient $\alpha_{bG}$ for a concrete bond failure lies between 0.50 and 0.85.

## 5.4 Shear Force Analyses

When analysing the shear capacity, the same requirements apply for near-surface-mounted CFRP strips as for externally bonded strips. This means that as described in Section 3.4.1, verifying the shear capacity should be carried out according to DIN EN 1992-1-1 [20] together with its associated National Annex [21]. As with externally bonded CFRP strips, the area of a near-surface-mounted strip may not be counted as part of the tension reinforcement $A_{sl}$ in Eq. (6.2a) of DIN EN 1992-1-1 [20]. Counting the CFRP strip as part of this reinforcement is not carried out in the DAfStb guideline because only a few shear tests have been carried out on strengthened members without shear reinforcement and so it is difficult to predict the effect of this. If the shear capacity analysis is not satisfied, shear strengthening for near-surface-mounted CFRP strips can be provided as described in Section 3.4.2.

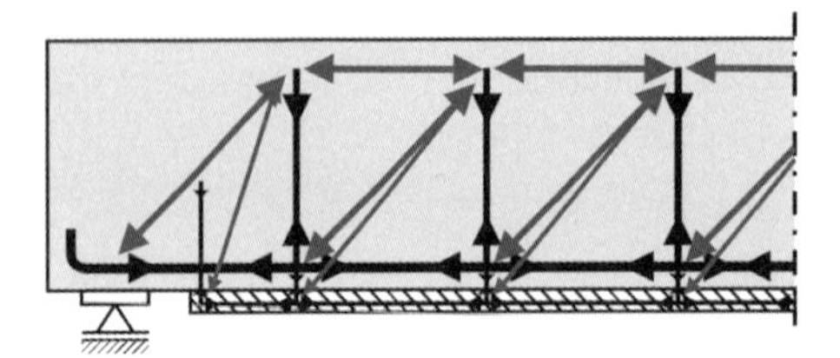

**Fig. 5.4** Mechanism for transferring tensile forces from externally bonded reinforcement to flexural compression zone of member by means of truss action

As with externally bonded CFRP strips, an analysis to prevent a concrete cover separation failure, see Section 3.4.3, is required for near-surface-mounted strips as well. Tests (see [11, 54]) have shown that the method described in Section 3.4.3 can also be used for members with near-surface-mounted CFRP strips.

In contrast to externally bonded CFRP strips, debonding at displaced crack edges does not occur with near-surface-mounted strips because the bond behaviour is much more robust. Therefore, the limit given in Section 3.4.1 for additional shear wrapping does not apply for near-surface-mounted CFRP strips. With very high shear loads, however, externally bonded shear straps must ensure that the tensile forces from the externally bonded reinforcement can also be tied back the flexural compression zone of the member with the help of truss action, as Figure 5.4 illustrates.

The limit value $\tau_{02}$ to DIN 1045 [94] has turned out to be a suitable variable (see [29]) for the maximum shear capacity without additional externally bonded shear straps. Equation 5.9 expresses this limit (see [11]):

$$V_{Ed} \leq 0.33 \cdot f_{ck}^{2/3} \cdot b_w \cdot d \tag{5.9}$$

If this limit value is exceeded, additional externally bonded shear straps are required to confine the strips.

## 5.5 Fatigue analysis

When checking fatigue for non-static loads, the DAfStb guideline can again be used to verify the bond of flexural strengthening in the form of near-surface-mounted CFRP strips. As the carbon fibres exhibit virtually no signs of fatigue, only the bond needs to be checked for fatigue when using CFRP strips. Besides the fatigue of the strengthening system, the concrete, reinforcing steel and prestressing steel must also be checked according to DIN EN 1992-1-1 [20] in conjunction with its National Annex [21].

In contrast to externally bonded CFRP strips, however, there is no comprehensive analysis concept available for near-surface-mounted strips. Owing to the low number of fatigue tests involving near-surface-mounted CFRP strips (see [27]), a quasi-fatigue strength analysis is the only option here. With so few test results available, it is not possible to specify an S-N curve for near-surface-mounted reinforcement. And as an S-N curve is unavailable, it is not possible to extrapolate for a number of load cycles greater than that given in the test results. Therefore, the analysis can only assume sufficient fatigue resistance for max. $2 \cdot 10^6$ load cycles. Design methods for numbers of load cycles $> 2 \cdot 10^6$ are not covered in the DAfStb guideline.

In this analysis, adequate resistance to fatigue for near-surface-mounted CFRP strips may be assumed for up to $2 \cdot 10^6$ load cycles provided the end anchorage force for a frequent cyclic action to DIN EN 1992-1-1 section 6.8.3 (3), and taking into account the 'shift rule', does not exceed the value $0.6F_{bLRd}$ ($F_{bLRd}$ to Equation 5.5) and the strip stress range does not exceed a value given by Equation 5.10. The strip thickness $t_L$ in mm should be used here so that the result is an admissible stress range in N/mm$^2$.

$$\Delta\sigma_L \leq \frac{500\,\mathrm{N/mm^2}}{t_L} \tag{5.10}$$

## 5.6 Analyses for the serviceability limit state

The analyses for the serviceability limit state, which were described for externally bonded CFRP strips in Section 3.6, also apply correspondingly for near-surface-mounted CFRP strips.

It should be pointed out here that owing to their effective and relatively stiff bond behaviour (see Figure 5.1), near-surface-mounted CFRP strips are ideal for retrofitting to control crack widths (see [109], for example). The method for allowing for the crack-limiting effect of near-surface-mounted reinforcement is based on a method proposed in [91], which assumes a bond-related interaction between the internal reinforcement and the near-surface-mounted reinforcement. It is assumed here that the cracks are closed or grouted at the time of strengthening and therefore no significant action effects due to residual stresses and loads are present. In this method it is first necessary to calculate the strip stress due to the load or restraint and assume a crack width. Owing to the assumed crack width, Equation 5.11 can be used to calculate the slip of the internal reinforcement and the near-surface-mounted reinforcement:

$$w_k = 2 \cdot s_{sr} = 2 \cdot s_{Lr} \tag{5.11}$$

With the help of the slip it is now possible to determine the mean bond stresses, the crack spacing and the mean strains using the equations given in the DAfStb guideline. The crack width can then be calculated with Equation 5.12:

$$w_k = s_{cr,max} \cdot (\varepsilon_{Lm} - \varepsilon_{cm}) \tag{5.12}$$

If the crack width from Equation 5.11 agrees with the assumption in Equation 5.12, this is the crack width that will occur.

## 5.7 Detailing

Essentially, near-surface-mounted CFRP strips must comply with the same detailing rules as those for externally bonded strips, which are described in Section 3.7. However, when it comes to the strip spacing, near-surface-mounted CFRP strips must comply not only with a maximum spacing, which is dealt with in the DAfStb guideline in the same way as the externally bonded CFRP strips, but also with a minimum spacing. Further to this minimum spacing there are also enhanced requirements regarding the distance of a strip from the edge of a member.

The DAfStb guideline specifies the minimum spacing $a_L$ for near-surface-mounted CFRP strips by way of Equation 5.13, which is based on the diameter $\phi$ of the steel reinforcing bars running parallel to the CFRP strips, the clear spacing $a_s$ of these steel reinforcing bars, the maximum aggregate size $d_g$ and the strip width $b_L$:

$$a_L \geq \begin{cases} d_g & \text{for } a_s \leq 2 \cdot \phi \\ b_L & \text{for } a_s > 2 \cdot \phi \end{cases} \tag{5.13}$$

This minimum spacing is necessary because where individual near-surface-mounted CFRP strips are too close together, one conceivable failure mode involves the strips together with the concrete cover become fully detached from the member (see [27]). The DAfStb guideline therefore includes the rules of [29], which are based on similar rules for internal steel reinforcing bars according to [27]. The final criterion for a minimum distance between CFRP strips is guaranteeing being able to cut the slots without damaging the member, which is also the case with the limits specified above.

A minimum edge distance is necessary because of the risk that the edge of the concrete member could break away if the spacing between a CFRP strip and the free edge of a member is too small and also the risk of damage to the edge of the concrete when cutting the slots. This minimum edge distance is specified in the DAfStb guideline by way of Equation 5.14. This approach was in the detailing rules of an earlier approval [29] and is based on [27].

$$a_r \geq \max \begin{cases} d_g \\ 2 \cdot b_L \end{cases} \tag{5.14}$$

The DAfStb guideline contains another requirement regarding the edge distance for the case where CFRP strips are being bonded to the soffit and the side face at the same time. This is because strips meeting along an edge cause a higher stress in the concrete at this corner.

# 6 Example 2: Strengthening a beam with near-surface-mounted CFRP strips

## 6.1 System

### 6.1.1 General

Owing to a change of use for a single-storey shed, a reinforced concrete downstand beam must carry higher loads and therefore needs to be strengthened. As-built documents with structural calculations to DIN 1045 [94] are available. The downstand beam, which was designed as a simply supported member, is to be strengthened with near-surface-mounted CFRP strips. It is assumed that the beam is free to rotate at its supports. Moderately damp conditions prevail in the building and the loads are primarily static. Figure 6.1 shows the structural system requiring strengthening and Figure 6.2 shows an idealized section through the beam.

### 6.1.2 Loading

The loads are predominantly static. Three load cases will be investigated for ultimate limit state design:

- **Load case 1** represents the situation prior to strengthening.
- **Load case 2** is the loading during strengthening. The strengthening measures are carried out under the dead load of the beam. Existing fitting-out items will be removed during the strengthening work.
- **Load case 3** represents the loading situation in the strengthened condition.

Table 6.1 lists the actions of the various load cases for the loads given in Figure 6.1.

Load case 3 governs for designing the strengthening measures. The load combination for the ultimate limit state and the load combination for the serviceability limit state under a rare load combination are required for the analyses. These load combinations are given by DIN EN 1990 [24] together with its associated National Annex [25]. The following applies for the ultimate limit state (persistent and transient design situations):

$$\sum_{j\geq 1} \gamma_{G,j} \cdot G_{k,j} + \gamma_P \cdot P + \gamma_{Q,1} \cdot Q_{k,1} + \sum_{i>1} \gamma_{Q,i} \cdot \psi_{0,i} \cdot Q_{k,i}$$

$$p_d = \gamma_G \cdot (g_{1,k} + g_{2,k}) + \gamma_Q \cdot q_k = 1.35 \cdot (30 + 5) + 1.5 \cdot 5.0 = 122.35\,\text{kN/m}$$

The load for the serviceability limit state is calculated as follows for a rare load combination:

$$\sum_{j\geq 1} G_{k,j} + P + Q_{k,1} + \sum_{i>1} \psi_{0,i} \cdot Q_{k,i}$$

*Strengthening of Concrete Structures with Adhesively Bonded Reinforcement: Design and Dimensioning of CFRP Laminates and Steel Plates*. First Edition. Konrad Zilch, Roland Niedermeier, and Wolfgang Finckh.

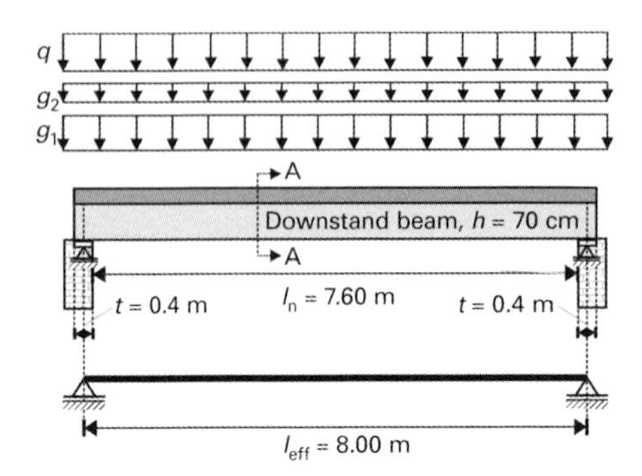


**Fig. 6.1** Downstand beam system requiring strengthening

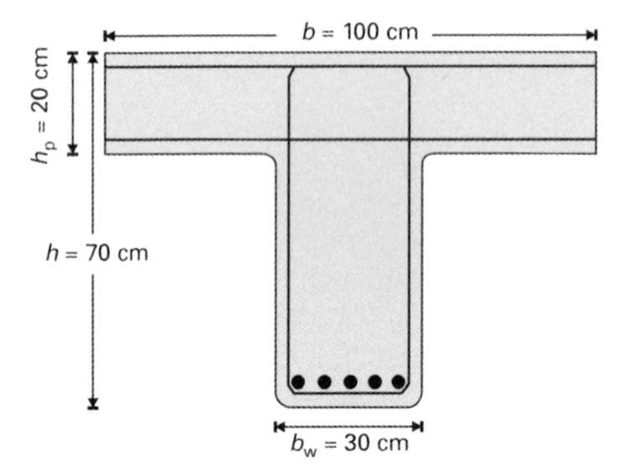


**Fig. 6.2** Section through downstand beam, (section A-A)

$$p_{\text{rare}} = g_{1,\text{k}} + g_{2,\text{k}} + q_{\text{k}} = 30 + 5 + 50 = 85\,\text{kN/m}$$

In order to determine the prestrain condition during strengthening, which according to DAfStb guideline [1, 2] part 1 section 5.1.1 (RV 19) must be considered for a quasi-permanent load combination, we get the following for load case 2:

$$\sum_{j\geq 1} G_{\text{k,j}} + P + \sum_{i\geq 1} \psi_{2,\text{i}} \cdot Q_{\text{k,i}}$$

$$p_{\text{perm}} = g_{1,\text{k}} = 30\,\text{kN/m}$$

**Table 6.1** Loads on the system in kN/m$^2$ for the various load cases.

| Load case | 1 | 2 | 3 |
|---|---|---|---|
| $g_{1,\text{k}}$ (dead load) | 30.0 | 30.0 | 30.0 |
| $g_{2,\text{k}}$ (fitting-out load) | 5.0 | — | 5.0 |
| $q_{\text{k}}$ (imposed load, category B) | 25.0 | — | 50.0 |

### 6.1.3 Construction materials

#### 6.1.3.1 Concrete compressive strength

Concrete of class B35 was able to be ascertained from the as-built documents according to DIN 1045 [94]. Following a test on the member, the result was strength class C30/37. Therefore, the values according to DIN EN 1992-1-1 [20] Tab. 3.1 for C30/37 concrete will be used for the design. This results in a mean concrete compressive strength $f_{cm} = 38\,\text{N/mm}^2$ and a characteristic concrete compressive strength $f_{ck} = 30\,\text{N/mm}^2$.

#### 6.1.3.2 Type and quantity of existing reinforcement

According to the as-built documents, the longitudinal reinforcement is five Ø28 mm ribbed steel reinforcing bars ($A_{sl} = 30.79\,\text{cm}^2$) and shear reinforcement in the form of vertical Ø8 mm links @ 200 mm c/c ($A_{sw}/s = 5.03\,\text{cm}^2$). It is apparent from the documents that the reinforcing steel is grade BSt 500 S (IV S) to [94] or [97]. Consequently, we can assume a yield stress $f_{syk} = 500\,\text{N/mm}^2$ and a modulus of elasticity $E_s = 200\,\text{kN/mm}^2$.

#### 6.1.3.3 Position of existing reinforcement

The as-built documents indicate a concrete cover of min $c = 2.0\,\text{cm}$, or nom $c = 3.0\,\text{cm}$, according to DIN 1045 [94]. A survey according to [98] has revealed that the reinforcement is positioned as shown in Figure 6.3.

#### 6.1.3.4 Strengthening system

Commercially available CFRP strips with a characteristic tensile strength $f_{Luk} = 2400\,\text{N/mm}^2$ and modulus of elasticity $E_L = 170\,\text{kN/mm}^2$ are to be bonded in slots for the strengthening. Strips with dimensions of ($t_L \times b_L$) 20 × 2 mm are to be used. The system includes an appropriate epoxy resin adhesive, for which a tensile strength $f_{Gtk} = 30\,\text{N/mm}^2$ and a compressive strength $f_{Gck} = 90\,\text{N/mm}^2$ will be assumed in the design. The other coefficients specific to this system are $k_{sys} = 0.8$, $k_{bck} = 2.5$, $\alpha_{bc} = 0.9$ and $\alpha_{bG} = 0.5$.

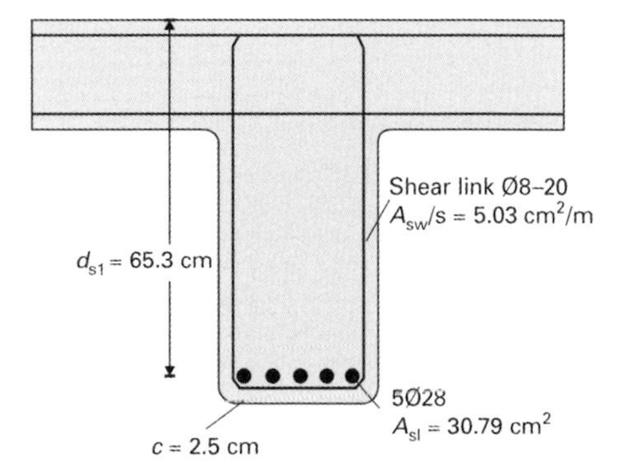


**Fig. 6.3** Type and position of existing reinforcement.(other reinforcement omitted for clarity)

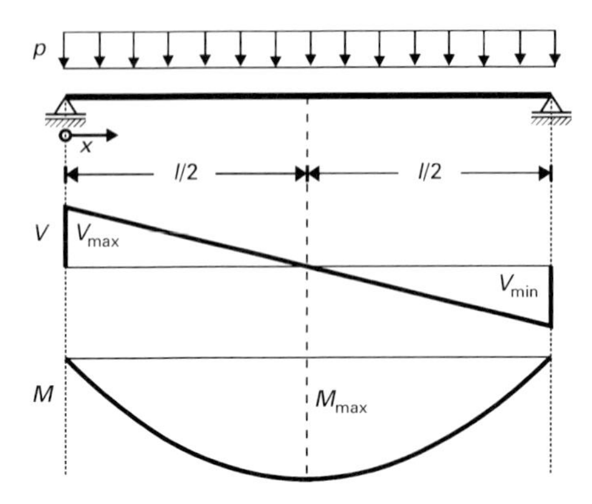


**Fig. 6.4** Shear forces and bending moments

## 6.2 Internal forces

Figure 6.4 shows the basic bending moment and shear force diagrams for the simply supported beam. The actual maximum values for the load combinations relevant to the design are given in Table 6.2.

$$M(x) = \frac{p}{2} \cdot l \cdot x - \frac{p \cdot x^2}{2}$$

$$V(x) = \frac{p}{2} \cdot l - p \cdot x$$

## 6.3 Determining the prestrain

DAfStb guideline [1, 2] part 1 section 5.1.1 (RV 19) requires that the prestrain be taken into account in the design. This is determined below using the example of the maximum moment. As according to the DAfStb guideline a prestrain should be determined with a quasi-permanent load combination for the serviceability limit state, characteristic material parameters are used in this section.

**Table 6.2** Maximum shear forces and bending moments for the relevant load combinations.

| Load combination | $M_{max}$ | $V_{max}$ | $V_{min}$ |
|---|---|---|---|
| — | kNm | kN | kN |
| Load case 3; ULS | 978 | 489 | –489 |
| Load case 3; SLS, rare | 680 | 340 | –340 |
| Load case 2; SLS, quasi-permanent | 240 | 120 | –120 |

An iterative method is used to determine the prestrain condition in the cross-section. The calculation below uses the internal lever arm of the reinforcing steel, determined iteratively, in order to demonstrate the method briefly. The internal lever arm, which represents the iteration variable, is

$$z_{s1} \approx 0.905 \cdot d_{s1} \approx 0.904 \cdot 653 \approx 590.4 \text{ mm}$$

The tensile force in the steel at the time of strengthening for the maximum moment can be calculated from the moment and the internal lever arm (see Section 3.2 and Figure 3.3):

$$F_{s1} = \frac{M_{0,k}}{z_{s1}} = \frac{240 \cdot 10^6}{590.4} = 406.5 \text{ kN}$$

Following on from that it is possible to determine the prestrain in the reinforcing steel from the area of the reinforcing bars and the modulus of elasticity of the reinforcement:

$$\varepsilon_{s1} = \frac{F_{s1}}{A_{s1} \cdot E_s} = \frac{406.5 \cdot 10^3}{30.79 \cdot 10^2 \cdot 200} = 0.66 \text{ mm/m}$$

Assuming a compressive strain in the concrete $\varepsilon_c > -2$ mm/m and a compression zone contained completely within the slab, the compressive force in the concrete according to Section 3.2 can be calculated approximately using the parabola-rectangle diagram for concrete under compression as follows:

$$F_c = b \cdot x \cdot f_{ck} \cdot \alpha_R = b \cdot \xi \cdot d_{s1} \cdot f_{ck} \cdot \left(-\frac{\varepsilon_c^2}{12} - \frac{\varepsilon_c}{2}\right)$$

$$= 1000 \cdot \left(\frac{-\varepsilon_c}{-\varepsilon_c + \varepsilon_{s1}}\right) \cdot 653 \cdot 30 \cdot \left(-\frac{\varepsilon_c^2}{12} - \frac{\varepsilon_c}{2}\right)$$

Equilibrium of the internal forces results in an equation for calculating the compressive strain in the concrete:

$$F_{s1} = F_c$$

$$406.5 \text{ kN} = -1000 \cdot \left(\frac{-\varepsilon_c}{-\varepsilon_c + 0.66}\right) \cdot 653 \cdot 30 \cdot \left(-\frac{\varepsilon_c^2}{12} - \frac{\varepsilon_c}{2}\right)$$

Solving the equation results in $\varepsilon_c = -0.26$ mm/m. As this value is $> -2$ mm/m, the above assumption was justified. The relative depth of the compression zone $\xi$ and the depth of the compression zone $x$ can now be determined with the help of the strains. As the depth of the compression zone is less than the depth of the slab, the above assumption – compression zone located fully within slab – was correct.

$$\xi = \frac{-\varepsilon_c}{-\varepsilon_c + \varepsilon_s} = \frac{0.26}{0.26 + 0.66} = 0.28$$

$$x = \xi \cdot d_{s1} = 0.28 \cdot 653 = 182.8 \text{ mm}$$

Using the coefficient $k_a$ (for $\varepsilon_c > -2$ mm/m), calculated according to Section 3.2, it is now possible to determine the internal lever arm $z_{s1}$:

$$k_a = \frac{8 + \varepsilon_c}{24 + 4 \cdot \varepsilon_c} = \frac{8 - 0.26}{24 - 4 \cdot 0.26} = 0.34$$

$$a = k_a \cdot \xi \cdot d_{s1} = 0.34 \cdot 0.28 \cdot 653 = 62.6 \text{ mm}$$

$$z_{s1} = d_{s1} - a = 653 - 62.6 = 590.4 \text{ mm}$$

As the internal lever arm roughly corresponds to the assumed lever arm, the resistance of the reinforced concrete cross-section at the position of the acting moment is

$$M_{Rk,0} = z_{s1} \cdot F_{s1} = 590.4 \cdot 406.5 \cdot 10^{-3} = 240 \text{ kNm}$$

The prestrain for the concrete therefore amounts to $\varepsilon_{c,0} = -0.26$ mm/m, and for the reinforcing steel $\varepsilon_{s1,0} = 0.66$ mm/m.

## 6.4 Verification of flexural strength

In the following calculations it is assumed that five strips are required for strengthening. The total strip cross-section is therefore

$$A_L = n_L \cdot t_L \cdot b_L = 5 \cdot 2 \cdot 20 = 200 \text{ mm}^2$$

When strengthening a member by means of near-surface-mounted CFRP strips, the slot dimensions must satisfy certain requirements, which influence the effective structural depth of the strips. According to DAfStb guideline [1, 2] part 3 section 4.4.1 (3), the depth of each slot in the concrete is

$$t_s \leq c - \Delta c_{dev}$$

The allowance $\Delta c_{dev}$ is made up as follows according to Section 5.2:

$$\Delta c_{dev} = \Delta c_{tool} + \Delta c_{slot} + \Delta c_{member} = 1 + 2 + 2 = 5 \text{ mm}$$

With a concrete cover $c = 25$ mm, the ensuing slot depth is $t_s = 20$ mm, which is exactly the same as the strip width $b_L$. The effective structural depth of the CFRP strip according to DAfStb guideline [1, 2] part 1 Eq. (RV 6.53) depends on the depth of the slot and is

$$d_L = h - \left(t_s - \frac{b_L}{2}\right) = 700 - \left(20 - \frac{20}{2}\right) = 690 \text{ mm}$$

The maximum strain that may be assumed in the design is determined from DAfStb guideline [1, 2] part 1 Eq. (RV 6.52) using the characteristic tensile strength of the strip $f_{Luk}$, the safety factor for strip failure $\gamma_{LL}$ and the coefficient $k_\varepsilon$:

$$\varepsilon_{LRd,max} = \kappa_\varepsilon \cdot \varepsilon_{Lud} = \kappa_\varepsilon \cdot \frac{f_{Luk}}{\gamma_{LL} \cdot E_L} = 0.8 \cdot \frac{2400}{1.2 \cdot 170 \cdot 10^3} = 9.41 \text{ mm/m}$$

The flexural strength is checked at mid-span for the maximum moment. In the following calculations it is assumed that the maximum strain in the strip can be exploited. As the strain in the strip $\varepsilon_{LRd,max} > f_{yd}/E_s$, we shall continue to assume that the reinforcing steel is yielding. Therefore, the tensile force in the reinforcing steel and the tensile force in the externally bonded reinforcement are

$$F_{s1d} = \frac{A_{s1} \cdot f_{yk}}{\gamma_s} = \frac{30.79 \cdot 10^2 \cdot 500}{1.15} = 1338.6\,\text{kN}$$

$$F_{LRd} = \varepsilon_{LRd,max} \cdot A_L \cdot E_L = 9.41 \cdot 200 \cdot 170 \cdot 10^3 = 320.0\,\text{kN}$$

The prestrain at the level of the near-surface-mounted CFRP strips is calculated using the prestrain in the reinforcement steel determined in Section 6.3:

$$\varepsilon_{L,0} = \varepsilon_{s1,0} + \frac{d_L - d_{s1}}{d_{s1}} \cdot \left(\varepsilon_{s1,0} + \varepsilon_{c,0}\right) = 0.66 + \frac{690 - 653}{653} \cdot (0.66 + 0.26) = 0.71\,\text{mm/m}$$

The total strain in the cross-section at the level of the strips is therefore

$$\varepsilon_{L,0} + \varepsilon_{LRd,max} = 0.71 + 9.41 = 10.12\,\text{mm/m}$$

Assuming a compressive strain in the concrete $\varepsilon_c < -2$ mm/m and that the compression zone is contained completely within the slab, the compressive force in the concrete can be expressed as follows according to Section 3.2:

$$F_{cd} = b \cdot x \cdot f_{cd} \cdot \alpha_R = b \cdot \xi \cdot d_L \cdot f_{ck} \cdot \frac{\alpha_{cc}}{\gamma_c} \cdot \left(1 + \frac{2}{3 \cdot \varepsilon_c}\right)$$

$$= 1000 \cdot \left(\frac{-\varepsilon_c}{-\varepsilon_c + \varepsilon_{L,0} + \varepsilon_{LRd,max}}\right) \cdot 690 \cdot 30 \cdot \frac{0.85}{1.5} \cdot \left(1 + \frac{2}{3 \cdot \varepsilon_c}\right)$$

Equilibrium of the internal forces enables the strain in the concrete to be subsequently calculated:

$$F_{s1d} + F_{Ld} = F_{cd}$$

Iteration results in $\varepsilon_c = -2.47$ mm/m. As this value is greater than the maximum compressive strain in the concrete $\varepsilon_{cu} = -3.5$ mm/m and also less than $\varepsilon_c = -2$ mm/m, the above assumption was justified. The relative depth of the compression zone $\xi$ and the depth of the compression zone $x$ can now be determined with the help of the strains. As the depth of the compression zone is less than the depth of the slab, the above assumption – compression zone located fully within slab – was correct.

$$\xi = \frac{-\varepsilon_c}{-\varepsilon_c + \varepsilon_{L,0} + \varepsilon_L} = \frac{2.47}{2.47 + 0.71 + 9.41} = 0.196$$

$$x = \xi \cdot d_L = 0.196 \cdot 690 = 135.4\,\text{mm}$$

Using the coefficient $k_a$ (for $\varepsilon_c < -2$ mm/m), which is the result according to Section 3.2, it is now possible to determine the internal lever arms:

$$k_a = \frac{3 \cdot \varepsilon_c^2 + 4 \cdot \varepsilon_c + 2}{6 \cdot \varepsilon_c^2 + 4 \cdot \varepsilon_c} = \frac{3 \cdot 2.47^2 - 4 \cdot 2.47 + 2}{6 \cdot 2.47^2 - 4 \cdot 2.47} = 0.39$$

$$a = k_a \cdot \xi \cdot d_L = 0.39 \cdot 0.196 \cdot 690 = 53.0 \text{ mm}$$

$$z_{s1} = d_{s1} - a = 653 - 53.0 = 600.0 \text{ mm}$$

$$z_L = h - a = 690 - 53.0 = 637.0 \text{ mm}$$

The moment capacity of the strengthened reinforced concrete cross-section is therefore

$$M_{Rd} = z_{s1} \cdot F_{s1d} + z_L \cdot F_{LRdL} = (1338.6 \cdot 600 \cdot 10^{-3} + 320 \cdot 637 \cdot 10^{-6}) = 1006.9 \text{ kNm}$$

As the moment capacity is greater than the acting moment of 978 kNm, the design is verified.

## 6.5 Bond analysis

### 6.5.1 Analysis point

According to DAfStb guideline [1, 2] part 1, RV 6.1.3.3 (RV 2), or Fig. RV 6.12, the analysis should be carried out, as described in section 5.3, at the point at which the CFRP strip is first required for loadbearing purposes. To do this we determine the point on the unstrengthened member at which the existing reinforcing steel reaches its yield point under the loads in the strengthened condition (load case 3). So we must first determine the bending moment at which the reinforcing steel begins to yield. The tensile force and the strain in the reinforcing steel for this situation are

$$F_{s1d} = \frac{A_{s1} \cdot f_{yk}}{\gamma_s} = \frac{30.79 \cdot 10^2 \cdot 500}{1.15} = 1338.6 \text{ kN}$$

$$\varepsilon_{s1} = \frac{f_{yd}}{E_s} = \frac{435}{200\,000} = 2.175 \text{ mm/m}$$

Assuming a compressive strain in the concrete $\varepsilon_c > -2$ mm/m and a compression zone contained completely within the slab, the compressive force in the concrete can be expressed as follows according to Section 3.2:

$$F_c = b \cdot x \cdot f_{ck} \cdot \alpha_R = b \cdot \xi \cdot d_{s1} \cdot f_{cd} \cdot \left( -\frac{\varepsilon_c^2}{12} - \frac{\varepsilon_c}{2} \right)$$

$$= 1000 \cdot \left( \frac{-\varepsilon_c}{-\varepsilon_c + \varepsilon_{s1}} \right) \cdot 653 \cdot 30 \cdot \frac{0.85}{1.5} \cdot \left( -\frac{\varepsilon_c^2}{12} - \frac{\varepsilon_c}{2} \right)$$

Equilibrium of the internal forces enables the strain in the concrete to be subsequently calculated:

$$F_{s1} = F_c$$

$$1338.6\,\text{kN} = -1000 \cdot \left(\frac{-\varepsilon_c}{-\varepsilon_c + 2.175}\right) \cdot 653 \cdot 30 \cdot \frac{0.85}{1.5} \cdot \left(-\frac{\varepsilon_c^2}{12} - \frac{\varepsilon_c}{2}\right)$$

Solving the equation results in $\varepsilon_c = -0.94$ mm/m. The relative depth of the compression zone $\xi$ and the depth of the compression zone $x$ can now be determined with the help of the strains. As the depth of the compression zone is less than the depth of the slab, the above assumption – compression zone located fully within slab – was correct.

$$\xi = \frac{-\varepsilon_c}{-\varepsilon_c + \varepsilon_s} = \frac{0.94}{0.94 + 2.175} = 0.30$$

$$x = \xi \cdot d_{s1} = 0.30 \cdot 653 = 195.9\,\text{mm}$$

Using the coefficient $k_a$ (for $\varepsilon_c > -2$ mm/m), i.e. the result according to Section 3.2, it is now possible to determine the internal lever arm $z_{s1}$:

$$k_a = \frac{8 + \varepsilon_c}{24 + 4 \cdot \varepsilon_c} = \frac{8 - 0.94}{24 - 4 \cdot 0.94} = 0.35$$

$$a = k_a \cdot \xi \cdot d_{s1} = 0.35 \cdot 0.30 \cdot 653 = 68.6\,\text{mm}$$

$$z_{s1} = d_{s1} - a = 653 - 68.6 = 584.4\,\text{mm}$$

The moment at which the reinforcing steel begins to yield is therefore

$$M_{\text{Rdy},0} = z_{s1} \cdot F_{s1} = 584.4 \cdot 1338.6 = 780.3\,\text{kNm}$$

The point at which the existing steel reinforcement reaches its yield point under the loads in the strengthened condition (load case 3) is found by solving the parabolic moment equation of Section 6.2:

$$x\left(M_{\text{Rdy},0}\right) = \frac{1}{2} - \sqrt{\frac{1^2}{4} - 2 \cdot \frac{M_{\text{Rdy},0}}{p_d}} = \frac{8}{2} - \sqrt{\frac{8^2}{4} - 2 \cdot \frac{780.3}{122.35}} = 2.20\,\text{m}$$

According to DAfStb guideline [1, 2] part 1, RV 6.1.3.3 (RV 2), or Fig. RV 6.12, the analysis point should be determined taking into account the shifted tensile force envelope. The 'shift rule' is calculated according to DIN EN 1992-1-1 section 9.2.1.3:

$$a_l = z \cdot (\cot\theta - \cot\alpha)/2 = 0.9 \cdot 656 \cdot (1.67 - 0)/2 = 491.8\,\text{mm}$$

The angle of the strut for the shear design is taken here from Section 6.6. The analysis point is therefore found to be at $x = 1.71$ m.

### 6.5.2 Acting strip force

As considering the prestrain in the bond analysis leads to a lower bond stress, it is first necessary to check whether the prestrain can be included. The prestrain can be considered if the cross-section is already cracked at this point. As the actual member was not inspected, it is assumed in the following calculations that the cross-section is cracked, provided the quasi-permanent load prior to strengthening has caused cracks to form.

$$M_{\mathrm{LF1,perm}} \geq M_{\mathrm{cr}}$$

The quasi-permanent moment at the analysis point for load case 1 to which the unstrengthened cross-section was subjected – taking into account the ‘shift rule’ and with $\psi_2 = 0.3$ to DIN EN 1990 [24] and its associated National Annex [25] – is therefore

$$M_{\mathrm{LF1,perm}}(x = 1.71 + a_l = 2.2) = \frac{g_{1,k} + g_{2,k} + \psi_2 \cdot q_k}{2} \cdot 1 \cdot x - \frac{(g_{1,k} + g_{2,k} + \psi_2 \cdot q_k) \cdot x^2}{2} =$$
$$= \frac{30 + 5 + 0.3 \cdot 25}{2} \cdot 8.0 \cdot 2.2 - \frac{(30 + 5 + 0.3 \cdot 25) \cdot 2.2^2}{2} = 271.15\,\mathrm{kNm}$$

The cracking moment for the cross-section can be calculated, for example, according to DAfStb guideline [1, 2] part 1, RV 6.1.1.3.3 Eq. (RV 6.5), as described in Section 3.3.3.2:

$$M_{\mathrm{cr}} = \kappa_{\mathrm{fl}} \cdot f_{\mathrm{ctm}} \cdot W_{\mathrm{c,0}} = 1.0 \cdot 2.9 \cdot 31.8 = 92.2\,\mathrm{kNm}$$

In this calculation the tensile strength of the concrete was taken from DIN EN 1992-1-1 Tab. 3.1 and the section modulus calculated as $W_{c,0} = 31.8 \cdot 10^6\,\mathrm{mm}^3$. The moment under quasi-permanent loading prior to strengthening is greater than the cracking moment and so it is assumed that the cross-section is already cracked.

$$M_{\mathrm{LF1,perm}} = 21.21\,\mathrm{kNm/m} < M_{\mathrm{cr}} = 29.87\,\mathrm{kNm/m}$$

The force in the strip taking into account the prestrain and the ‘shift rule’ is calculated below. Table 6.3 lists the strains and internal forces at this point.

**Table 6.3** Strains and internal forces at bond analysis point.

| $x$ | $M_{\mathrm{Ed}}$ | $\varepsilon_{s,0}$ | $\varepsilon_{c,0}$ | $\varepsilon_L$ | $\varepsilon_s$ | $\varepsilon_c$ | $F_{\mathrm{LEd}}$ | $F_{\mathrm{sEd}}$ | $F_{\mathrm{cEd}}$ |
|---|---|---|---|---|---|---|---|---|---|
| m | kNm | mm/m | mm/m | mm/m | mm/m | mm/m | kN | kN | kN |
| 2.2 | 780.3 | 0.48 | −0.19 | 1.76 | 2.10 | −0.93 | 59.77 | 1294.84 | −1354.55 |

### 6.5.3 Bond resistance

First of all, the bond length of the near-surface-mounted CFRP strip is required to determine the bond resistance. The bond length is the result of the analysis point in Section 6.5.1 minus the distance of the strip from the centre of the support. To make it easier to cut the slot, the distance of the strip from the edge of the support is specified as 200 mm. According to Figure 6.1, the distance from the edge of the support to the centre of the support is another 200 mm. The bond length available is therefore

$$l_{bL} = x - a_L = 1710 - 200 - 200 = 1310\,\text{mm}$$

To determine the bond strength, the maximum bond stress in the adhesive and the maximum bond stress in the concrete are required according to DAfStb guideline [1, 2] part 1 Eqs. (RV 8.13) and (RV 8.14), using the variables from Section 6.1.3:

$$\tau_{bGk} = k_{sys} \cdot \sqrt{\left(2 \cdot f_{Gtk} - 2 \cdot \sqrt{\left(f_{Gtk}^2 + f_{Gck} \cdot f_{Gtk}\right)} + f_{Gck}\right) \cdot f_{Gtk}}$$

$$\tau_{bGk} = 0.8 \cdot \sqrt{\left(2 \cdot 30 - 2 \cdot \sqrt{\left(30^2 + 90 \cdot 30\right)} + 90\right) \cdot 30} = 24\,\text{N/mm}^2$$

$$\tau_{bck} = k_{bck} \cdot \sqrt{\text{f}_{cm}} = 4.5 \cdot \sqrt{38} = 27.7\,\text{N/mm}^2$$

The design value of the bond stress is now calculated with the long-term effect coefficients and the safety factor according to DAfStb guideline [1, 2] part 1 Eq. (RV 8.12):

$$\tau_{bLd} = \frac{1}{\gamma_{BE}} \cdot \min\begin{cases} \tau_{bGk} \cdot \alpha_{bG} \\ \tau_{bck} \cdot \alpha_{bc} \end{cases} = \frac{1}{1.3} \cdot \min\begin{cases} 24.0 \cdot 0,5 \\ 27.7 \cdot 0.85 \end{cases} = \frac{1}{1.3} \cdot 12 = 9.23\,\text{N/mm}^2$$

The tensile force per strip that can be anchored via the composite action between CFRP strip and concrete member can be calculated for $l_{bL} > 115$ mm to DAfStb guideline part 1 Eq. (RV 6.56):

$$F_{bLRd} = b_L \cdot \tau_{bLd} \cdot \sqrt[4]{a_r} \cdot \left(26.2 + 0.065 \cdot \tanh\left(\frac{a_r}{70}\right) \cdot (l_{bL} - 115)\right) \cdot 0.95$$

$$F_{bLRd} = 20 \cdot 9.23 \cdot \sqrt[4]{50} \cdot \left(26.2 + 0.065 \cdot \tanh\left(\frac{50}{70}\right) \cdot (1310 - 115)\right) \cdot 0.95 = 34.44\,\text{kN}$$

The edge distance of the strip $a_r$ here is such that it is also equal to the centre-to-centre spacing of the strips. The spacing and edge distance chosen in this way also comply with the requirement according to DAfStb guideline part 1, RV 8.2.1 (see Section 5.7 of this book).

$$a_{\mathrm{r}} = \frac{b_{\mathrm{w}}}{n_{\mathrm{L}} + 1} = \frac{300}{5 + 1} = 50\,\mathrm{mm}$$

The design value of the bond strength of all externally bonded reinforcement is obtained by multiplying the tensile force that can be anchored per strip by the number of strips. For simplicity, the most unfavourable edge distance of the outer strips was also applied to the other, inner, strips.

$$F_{\mathrm{bLRd,sum}} = n_{\mathrm{L}} \cdot F_{\mathrm{bLRd}} = 5 \cdot 34.44 = 172.22\,\mathrm{kN}$$

### 6.5.4 Bond analysis

The design value of the bond strength is greater than the acting strip force and so the bond analysis is regarded as verified:

$$F_{\mathrm{LEd}} = 59.77\,\mathrm{kN} \leq F_{\mathrm{bLRd,sum}} = 172.22\,\mathrm{kN}$$

## 6.6 Shear analyses

### 6.6.1 Shear capacity

First of all we shall attempt to analyse the shear capacity of the downstand beam according to DIN EN 1992-1-1 [20] and its associated National Annex [21]. Checking the capacity of the strut in the concrete is the first step. To do this, the design shear force is determined according to DIN EN 1992-1-1 [20, 21] section 6.2.1 (8):

$$V_{\mathrm{Ed,red,max}} = V_{\mathrm{Ed}} - p_{\mathrm{Ed}} \cdot a_i = 489.0 - 122.25 \cdot 0.20 = 464.6\,\mathrm{kN}$$

The maximum strut angle used in the design is obtained from DIN EN 1992-1-1 [20, 21] Eq. (6.7aDE):

$$1.0 \leq \cot\theta \leq \frac{1.2}{1 - V_{\mathrm{Rd,cc}}/V_{\mathrm{Ed}}} \leq 3.0$$

$$1.0 \leq \frac{1.2}{1 - 131.5/464.6} \leq 3.0 \Rightarrow \cot\theta = 1.67$$

The shear resistance $V_{\mathrm{Rd,cc}}$ to DIN EN 1992-1-1 [20, 21] Eq. (6.7bDE) is used here:

$$V_{\mathrm{Rd,cc}} = c \cdot 0.48 \cdot f_{\mathrm{ck}}^{1/3} \cdot b_{\mathrm{w}} \cdot z$$

$$V_{\mathrm{Rd,cc}} = 0.5 \cdot 0.48 \cdot 30^{1/3} \cdot 300 \cdot 0.9 \cdot 653 = 131.5\,\mathrm{kN}$$

The maximum shear resistance, which is limited by the strength of the strut, is calculated using DIN EN 1992-1-1 [20, 21] Eq. (6.9):

$$V_{\mathrm{Rd,max}} = \frac{\alpha_{\mathrm{cw}} \cdot b_{\mathrm{w}} \cdot z \cdot \nu_1 \cdot f_{\mathrm{cd}}}{\cot\theta + \tan\theta} = \frac{1.0 \cdot 300 \cdot 0.9 \cdot 653 \cdot 0.75 \cdot 17}{1.67 + 1/1.67} = 989.8\,\mathrm{kN}$$

The maximum shear resistance is greater than the design shear force and so the analysis of the strut in the concrete is verified.

$$V_{Rd,max} = 989.8\,kN \geq V_{Ed,red,max} = 464.6\,kN$$

When analysing the load-carrying capacity of the internal shear links, or rather the tie, the design shear force to DIN EN 1992-1-1 [20, 21] section 6.2.1 (8) may be taken as

$$V_{Ed,red,s} = V_{Ed} - p_{Ed} \cdot (a_i + d) = 489.0 - 122.25 \cdot (0.10 + 0.653) = 384.7\,kN$$

As a simplified approach, the analysis at this point uses the same strut angle as for the analysis of the strength of the strut in the concrete. When analysing the tie, the smaller strut angle leads to a lower load-carrying capacity, which therefore lies on the safe side. The shear resistance (limited by the yield stress of the shear reinforcement) is calculated using DIN EN 1992-1-1 [20, 21] Eq. (6.8).

$$V_{Rd,s} = \left(\frac{A_{sw}}{s}\right) \cdot z \cdot f_{ywd} \cdot \cot\theta = 0.503 \cdot 0.9 \cdot 653 \cdot 435 \cdot 1.67 = 215.0\,kN$$

The design shear force is greater than the resistance of the shear reinforcement, so the **analysis** of the tie is **not satisfied** and shear strengthening will be required.

$$V_{Rd,s} = 215.0\,kN \geq V_{Ed,red,s} = 384.7\,kN$$

### 6.6.2 Shear strengthening

Externally bonded full shear wrapping made from grade S235JR steel, nominal dimensions $t_{Lw} = 6$ mm and $b_{Lw} = 80$ mm at a centre-to-centre spacing $s_{Lw} = 600$ mm, will be used for the shear strengthening. The yield stress of grade S235JR steel according to DAfStb guideline [1, 2] part 2 is $f_{yk} = 0.8 \cdot 235\,N/mm^2 = 188\,N/mm^2$, and the modulus of elasticity $E_{Lw} = 200\,000\,N/mm^2$.

The additional shear force that can be accommodated is calculated according to DAfStb guideline [1, 2] part 1 Eq. (6.108):

$$V_{Rd,Lw} = \frac{A_{Lw}}{s_{Lw}} \cdot z \cdot f_{Lwd} \cdot \cot\theta$$

The area of shear strengthening is calculated according to DAfStb guideline [1, 2] part 1 Eq. (6.109):

$$\frac{A_{Lw}}{s_{Lw}} = \frac{2 \cdot t_{Lw} \cdot b_{Lw}}{s_{Lw}} = \frac{2 \cdot 6 \cdot 80}{600} = 1.6\,mm^2/mm$$

The capacity of the shear strengthening $f_{wLd}$ is determined depending on the material and the type of strengthening. As the downstand beam to be strengthened is a T-beam, only full wrapping is permitted according to DAfStb guideline part 1, RV 6.2.6 (RV 2). The strength of full wrapping in steel is the minimum of the yield stress and the stress that can be transferred across any laps:

$$f_{\mathrm{Lwd,GS}} = \min\{f_{\mathrm{yd}}; f_{\mathrm{Gud,Lw}}\}$$

A lap is planned on the soffit of the beam in accordance with DAfStb guideline [1, 2] part 1 Fig. RV 9.2. According to DAfStb guideline section RV 9.2.7.2 (RV 7), 260 mm is therefore available for this lap length. The maximum length of lap that can be counted according to DAfStb guideline part 1 Eq. (RV 6.112) is

$$l_{\mathrm{ü,max}} = 0.121 \cdot \sqrt{E_{\mathrm{Lm}} \cdot t_{\mathrm{L}}} = 0.121 \cdot \sqrt{200\,000 \cdot 6} = 132.6\,\mathrm{mm}$$

As $l_{\mathrm{ü,max}} < l_{\mathrm{ü}}$, the stress that can be transferred at the lap is calculated according to DAfStb guideline [1, 2] part 1 Eq. (RV 6.113):

$$f_{\mathrm{Lwd,GS}} = \frac{1.004}{\gamma_{\mathrm{BG}}} \cdot \sqrt{\frac{E_{\mathrm{L}}}{t_{\mathrm{Lw}}}} = \frac{1.004}{1.3} \cdot \sqrt{\frac{200\,000}{6}} = 141.0\,\mathrm{N/mm^2}$$

The strength of a steel shear strap to be used in the calculations is therefore

$$f_{\mathrm{Lwd,GS}} = \min\{f_{\mathrm{yd}}; f_{\mathrm{Gud,Lw}}\} = \min\{188; 141\} = 141.0\,\mathrm{N/mm^2}$$

The additional shear force that can be accommodated can be calculated using DAfStb guideline [1, 2] part 1 Eq. (RV 6.108):

$$V_{\mathrm{Rd,Lw}} = \frac{A_{\mathrm{Lw}}}{s_{\mathrm{Lw}}} \cdot z \cdot f_{\mathrm{Lwd}} \cdot \cot\theta = 1.6 \cdot 0.9 \cdot 653 \cdot 141 \cdot 1.67 = 221.42\,\mathrm{kN}$$

The total load-carrying capacity of the tie is therefore given by DAfStb guideline [1, 2] part 1 Eq. (RV 6.107):

$$V_{\mathrm{Rd}} = V_{\mathrm{Rd,s}} + V_{\mathrm{Rd,Lw}} = 215.0 + 221.4 = 436.4\,\mathrm{kN}$$

The load-carrying capacity of the tie is now greater than the design shear force and so the design with the shear strengthening is verified.

$$V_{\mathrm{Rd}} = 436.4\,\mathrm{kN} \geq V_{\mathrm{Ed,red,s}} = 384.7\,\mathrm{kN}$$

To complete the analysis, it is only necessary to check the fasteners for the steel which are required to anchor the shear straps in the compression zone (see Figure 3.10).

### 6.6.3 Check for concrete cover separation failure

When checking for a concrete cover separation failure, it is first necessary to calculate the shear resistance of a member without shear reinforcement. The shear resistance of a member without shear reinforcement is obtained from the maximum of Eqs. (6.2a) and (6.2b) in DIN EN 1992-1-1 [20, 21]. The design shear resistance according to Eq. (6.2a) is

$$V_{\mathrm{Rd,c}} = \left[C_{\mathrm{Rd,c}} \cdot k \cdot (100 \cdot \rho_{\mathrm{l}} \cdot f_{\mathrm{ck}})^{1/3} + 0.12 \cdot \sigma_{\mathrm{cp}}\right] \cdot d \cdot b_{\mathrm{w}}$$

The following shear resistance is calculated using the variables in Eq. (6.2a) according to DIN EN 1992-1-1 or its National Annex. It should be noted here that according to DAfStb guideline part 1 section 6.2.2 (RV 7) and DIN EN 1992-1-1 Fig. 6.3, the externally bonded reinforcement may not be counted as part of the longitudinal reinforcement.

$$k = 1 + \sqrt{\frac{200}{d}} = 1 + \sqrt{\frac{200}{653}} = 1.55 \leq 2.0$$

$$\sigma_{\mathrm{cp}} = N_{\mathrm{Ed}}/A_{\mathrm{c}} = 0$$

$$C_{\mathrm{Rd,c}} = \frac{0.15}{\gamma_{\mathrm{c}}} = \frac{0.15}{1.5} = 0.10$$

$$\rho_{\mathrm{l}} = \frac{A_{\mathrm{sl}}}{d \cdot b_{\mathrm{w}}} = \frac{3079}{653 \cdot 300} = 1.57\% \leq 2\%$$

$$V_{\mathrm{Rd,c}} = \left[0.10 \cdot 1.55 \cdot 1.0 \cdot (1.57 \cdot 30)^{1/3}\right] \cdot 653 \cdot 300 = 109.94\,\mathrm{kN}$$

The minimum shear resistance of a member without shear reinforcement is given by DIN EN 1992-1-1 Eq. (6.2b) as

$$V_{\mathrm{Rd,c}} = \left[\frac{0.0525}{\gamma_{\mathrm{c}}} \cdot \sqrt{k^3 \cdot f_{\mathrm{ck}}} + 0.12 \cdot \sigma_{\mathrm{cp}}\right] \cdot d \cdot b_{\mathrm{w}}$$

$$= \left[\frac{0.0525}{1.5} \cdot \sqrt{1.55^3 \cdot 30}\right] \cdot 653 \cdot 300 = 73.91\,\mathrm{kN}$$

The design shear resistance of this member without shear reinforcement is therefore $V_{\mathrm{Rd,c}} = 109.94$ kN.

The limit beyond which no shear wrapping at the end of the strip is necessary is calculated using DAfStb guideline [1, 2] part 1 Eq. (RV 6.121) depending on the distance of the strip from the centre of the support $a_{\mathrm{L}}$ according to Section 6.5.3:

$$V_{\mathrm{Rd,c,LE}} = 0.75 \cdot \left(1 + 19.6 \cdot \frac{(100 \cdot \rho_{\mathrm{sl}})^{0.15}}{a_{\mathrm{L}}^{0.36}}\right) \cdot V_{\mathrm{Rd,c}}$$

$$V_{\mathrm{Rd,c,LE}} = 0.75 \cdot \left(1 + 19.6 \cdot \frac{(1.57)^{0.15}}{400^{0.36}}\right) \cdot 109.94 = 282.52\,\mathrm{kN}$$

As the acting shear force is greater than the limit according to the DAfStb guideline, shear wrapping at the end of the strip is essential.

**Table 6.4** Strains and internal forces for determining force acting on end strap.

| $x$ | $\varepsilon_L$ | $\varepsilon_s$ | $\varepsilon_c$ | $F_{LEd}$ | $F_{sEd}$ | $F_{cEd}$ |
|---|---|---|---|---|---|---|
| m | mm/m | mm/m | mm/m | kN | kN | kN |
| 0.892 | 1.11 | 1.02 | −0.43 | 37.61 | 630.59 | −668.08 |

$$V_{Ed} = 489.0\,\text{kN} \leq V_{Rd,c,LE} = 282.52\,\text{kN/m}$$

The force acting on the end strap is calculated according to DAfStb guideline part 1 section RV 9.2.6:

$$F_{LwEd,end} = F^*_{LEd} \cdot \tan\theta = 37.61 \cdot \frac{1}{1.67} = 22.5\,\text{kN}$$

where $F^*_{LEd}$ is the fictitious strip tensile force at the end of the strip plus the 'shift rule'. This means that the strip force is required at the point $x = a_L + a_l = 400 + 491.8 = 891.8$ mm. This strip force and the associated strains are listed in Table 6.4 and were determined iteratively without taking the prestrain into account because this has a favourable effect here but it is not certain that the cross-section is cracked at this point.

The force acting on the end strap is carried by the end strap of the shear strengthening. For this reason, this strap will be somewhat wider. The additional width necessary is $b_{Lw} = 20$ mm and the additional resistance of the strap can be calculated with the following equation:

$$F_{LwRd,end} = 2 \cdot t_{Lw} \cdot b_{Lw} \cdot f_{Lwd} = 2 \cdot 6 \cdot 20 \cdot 141 = 33.84\,\text{kN}$$

The resistance is greater than the action of 22.5 kN and so the design is verified. To avoid a concrete cover separation failure, the end strap of the shear strengthening must therefore have dimensions of $(b_{Lw} \times t_{Lw})$ $100 \times 6$ mm.

## 6.7 Analyses for the serviceability limit state

Analyses of crack width and deformation are not carried out in this example. It is merely verified that the necessary stresses are complied with. According to DAfStb guideline part 1 section 7.2, described in Section 3.6 of this book, the strains in the strip and the reinforcing steel must be limited as follows for a rare load combination:

$$\varepsilon_s \leq \frac{f_{yk}}{E_s} = \frac{500}{200\,000} = 2.5\,\text{mm/m}$$

$$\varepsilon_L \leq 2\,\text{mm/m}$$

Under a rare load combination, we get the following maximum moment at mid-span:

$$M_{E,rare} = 680\,\text{kNm/m}$$

The prestrains $\varepsilon_{s,0} = 0.66$ mm/m, $\varepsilon_{c,0} = -0.26$ mm/m and $\varepsilon_{L,0} = 0.71$ mm/m are calculated as explained in Section 4.3. The strains $\varepsilon_L = 1.18$ mm/m, $\varepsilon_s = 1.76$ mm/m and $\varepsilon_c = -0.53$ mm/m were determined iteratively with the characteristic strengths and the following two conditions:

$$M_R = M_{E,rare}$$

$$F_{s1} + F_L = -F_c$$

As the ultimate strains for the strip and the reinforcing steel are not exceeded, the design is verified.

# 7 Design of column strengthening with CF sheets

## 7.1 Principles

As with other materials, triaxial compression loads on concrete lead to an increase in the compression that can be accommodated in the direction of the largest principal stress. Just a hydrostatic lateral pressure amounting to 20% of the uniaxial strength $f_{cm}$ of concrete results in a doubling of the admissible compressive stress; and the admissible deformations also increase considerably. In contrast to a specific load applied in the transverse direction, the effect of confining reinforcement resulting from the prevention of lateral strain is regarded as a passive lateral pressure. Owing to the large deformation capacity of the reinforcing steel, the normal situation in compression members with helical reinforcement, for example, is that the disintegration of the concrete microstructure leads to failure of the member (in a similar way to a triaxial compression test with hydrostatic lateral pressure). If the confining effect is achieved by including transverse reinforcement in the form of fibre-reinforced materials with a virtually linear elastic behaviour, then the lateral pressure rises continuously until the confining reinforcement fails in tension. Figure 7.1 shows a schematic representation of the effect of CFRP wrapping compared with a cross-section containing confining steel reinforcement and an unconfined section.

When it comes to describing the loadbearing behaviour numerically, a distinction has to be made between the load-carrying capacity of the cross-section, which essentially depends on the material properties and therefore can be described by tests (e.g. multi-axial compression tests) on small-format specimens, and the load-carrying capacity of the member, which besides the material properties is also dependent on the geometry of the member and the loading. Only in the case of a concentric load on a short column, in which the influence of slenderness can be excluded, is the load-carrying capacity of the cross-section equal to that of the member.

The development of the principles for designing confined concrete members is attributed to the French engineer *Armand Considére* [110, 111], who in 1902 patented a method for casting concrete elements with a high axial compressive strength. The particular feature of this method was that a metal helix, with closely spaced windings, was placed around the core of the concrete member. On the basis of his experimental studies, Considére formulated an initial addition function that considered the increase in the load-carrying capacity due to the confining reinforcement.

As early as 18 September 1909, the 'Circular decree concerning the design of concrete columns with confining iron bars' valid for the Kingdom of Prussia permitted an increase in the load due to the confining effect of helical transverse reinforcement according to Considére's method. The effect of confining reinforcement was subsequently described in numerous publications.

In the German language the studies by *Müller* [112] and *Menne* [113] are the most important. The design method in DIN 1045 (see [94], for example) for confined compression members was based on their investigations and remained valid and unchanged for more than 25 years. *Müller's* work was primarily based on tests on

*Strengthening of Concrete Structures with Adhesively Bonded Reinforcement: Design and Dimensioning of CFRP Laminates and Steel Plates*. First Edition. Konrad Zilch, Roland Niedermeier, and Wolfgang Finckh.

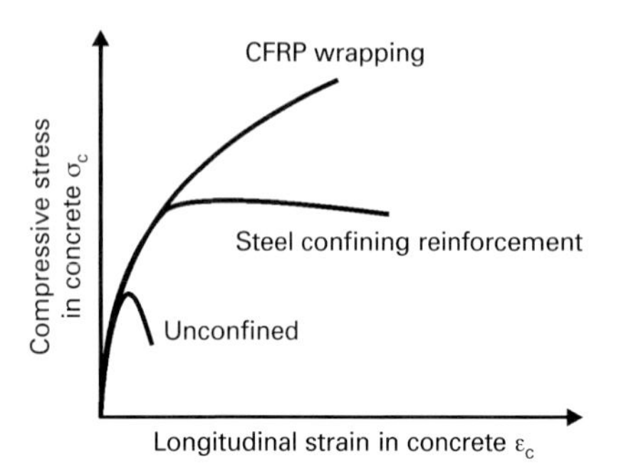


**Fig. 7.1** The effect of confining reinforcement

concentrically loaded, short (i.e. no or very little risk of buckling), confined columns with a circular cross-section. However, using the results of the few tests with eccentric loading available to him, he was already able to make a quantitative estimate of the reduction in the axial force capacity caused by simultaneous bending. In contrast to earlier editions of DIN 1045, the design concept based on *Müller's* work only allows eccentricities amounting to one-eighth of the core cross-section lying within the confining reinforcement, which means that the confining effect may only be assumed for columns compressed over their full cross-section.

*Müller* also made a proposal for ascertaining the influence of the flexural slenderness of the confined core which, however, did not find its way into any standard. It was not until the studies of *Menne* [113] demonstrated that confined columns with just moderate slenderness ratios suffered a considerable decline in their load-carrying capacity DIN 1045 [114] was updated accordingly in 1975 [115].

Redesigning to DIN EN 1992-1-1 [20, 21] allows higher loads to be assigned to reinforced concrete columns originally designed to DIN 1045 [94]. Therefore, in some circumstances it is possible to avoid costly, elaborate strengthening measures. This can be attributed to the lower partial safety factor for the portion of the load carried by the reinforcing steel and the higher permissible compressive strains in the concrete, which permit the reinforcing steel to be better utilized. Testing the concrete of the actual member to establish its strength is another way of possibly avoiding the need for strengthening measures. On the one hand, owing to the strength allowance and age hardening, concrete strengths are often higher than the values of the strength classes originally envisaged. On the other hand, knowledge of the material properties allows lower partial safety factors to be used on the resistance side of the equation (see [116, 117], for example).

However, if a structure is to undergo major changes, e.g. the addition of an extra storey, constructional measures to strengthen the vertical loadbearing members will be unavoidable in most cases. So far, the methods dealt with here, such as wrapping with fibre-reinforced materials, have not been covered by German standards or national technical approvals and could only be used on the basis of individual approvals for particular projects. In Germany, strengthening reinforced concrete columns on the basis of regulations approved by the building authorities could only be carried out using sprayed concrete designed according to DIN 18551 [118].

As for reinforced concrete columns strengthened with sprayed concrete neither the shell of sprayed concrete nor the additional reinforcement in that shell can be arranged with an interlocking structural connection, the load-carrying capacity in the load transfer regions must be guaranteed by the confining effect of the links in the original column and in the sprayed concrete. Owing to the requirements regarding the minimum distance of the reinforcing bars from the existing concrete and the concrete cover, a layer of sprayed concrete always results in a significant enlargement of the column cross-section. Strengthening with sprayed concrete is labour-intensive and has the disadvantages of dust, noise and moisture, and therefore should only be used when a substantial increase in a column's load-carrying capacity is necessary.

For reinforced concrete columns designed according to the earlier editions of DIN 1045 there are some strengthening assignments that fall midway between redesign and sprayed concrete. It is for these projects that wrapping with fibre-reinforced materials is to be recommended. This method has already been used frequently in Europe and elsewhere, often on the basis of design codes or guides covering the subsequent strengthening of reinforced concrete compression members by means of wrapping with fibre-reinforced materials. Such codes already exist in, for instance, the USA [119], the UK [120], Canada [121], Switzerland [122] and Italy [123]. However, it should be noted that none of these documents deal fully with the issues relevant to design. Instead, the analyses required are limited solely to the load-carrying capacity of the cross-section, which has already been the subject of an almost incalculable number of experimental studies involving small-format specimens. Other aspects, such as how the load-carrying capacity of the member differs from that of the cross-section, the deformation behaviour of the highly stressed concrete over time and the loadbearing behaviour of the fibre-reinforced materials depending on the duration of loading and the ambient conditions, are not addressed in these publications.

The objective of the subsequent confinement of a compression member is to increase either its load-carrying capacity or its deformation capacity. The latter is very important in countries where it is necessary to improve the seismic behaviour of members and structures that do not comply with the design codes based on the latest findings. In contrast to the strengthening of compression members, seismic loads mostly involve shear forces as well, which induce relatively high flexural stresses. For compression members designed to be concentrically loaded, flexural stresses are caused by an unintended load eccentricity, prescribed in the relevant standard, and second-order theory effects. These must be considered in the design in addition to axial forces and are intensified by the creep of the concrete. Owing to the minimal eccentricities, the members involved are therefore mostly loaded by predominantly axial loads. The applications covered by the DAfStb guideline [1, 2] are thus restricted to strengthening members loaded by axial loads with a small eccentricity, which corresponds to the projects encountered in Germany.

In accordance with the scope of the experimental findings to date, the intentional eccentricity in the DAfStb guideline is limited to one-quarter of the column diameter.

This restriction agrees with the recommendations of *Menne* [113] for reinforced concrete columns with helical reinforcement. His tests on members demonstrated that *Müller's* proposal [124] to limit the eccentricity to one-eighth of the core diameter, and which was included in the 1972 edition of DIN 1045 [114], was too conservative.

Further, the range of applications covered by the DAfStb guideline [1, 2] is limited to compression members with a circular cross-section. In principle, confinement can also increase the load-carrying capacity of members with a square or rectangular cross-section. However, the confinement can induce a sufficiently high transverse compressive stress in certain areas of the cross-section only. This fact has been verified by *Sheikh* and *Uzumeri* [125] for reinforced concrete columns. They showed that only the area lying inside parabolic arcs can be assumed to be effectively confined, where the arcs spring from the corners of the links. In square columns with typical dimensions, the ratio of the effective confined area to the area of the concrete is already <60%. As the cross-section deviates more and more from a square section, there is a rapid decrease in the effective confined area.

In addition, the high transverse compression acting on the fibre-reinforced material in the region of the rounded corner leads to a lower tensile stress that can be accommodated by the confining reinforcement and hence to a further drop in the effectiveness of the strengthening measures for rectangular cross-sections. On the whole it can be said that confinement in the form of fibre-reinforced materials for compression members with a rectangular cross-section does not represent a reasonable method of strengthening in most instances. Therefore, modifications to the method to allow for this type of cross-section have already been investigated in order to increase the efficiency. These modifications entail building up the cross-section into a circular or elliptical form (see [126], for example), or including expanding elements to prestress the wrapping between the rounded corners [127]. The latter function as additional supports for the arcs and hence lead to an increase in the effective confined area.

As the history of DIN 1045 has meant that high-strength concretes only began to be widely used in Germany quite recently, members made from normal-strength concrete represent the standard case for strengthening projects at the moment. The applications covered by the DAfStb guideline [1, 2] are therefore limited to the subsequent strengthening of existing members made from normal-strength concrete.

There are various ways of arranging externally bonded fibre-reinforced materials. The DAfStb guideline [1, 2] only deals with wrapping applied over the entire surface, which compared with wrapping applied in strips or in a spiral does not involve any reduction in the confining effect in the longitudinal direction of the column.

The range of applications covered by the DAfStb guideline [1, 2] is limited to carbon fibre (CF) sheets, which when used as confining reinforcement must have a national technical approval. The wet lay-up method is used to ensure that all fibres are fully soaked with resin.

The following sections explain the background to and sources of the provisions contained in the DAfStb guideline [1, 2]. A more detailed description of the underlying models can be found in [56] and the references given in the following sections. As at the time of preparing the guideline there were no national technical approvals available for CF sheets for strengthening reinforced concrete columns, but the industry representatives on the DAfStb subcommittee responsible for drawing up the guideline wanted expressly to include design concepts for column strengthening, the formulations in the guideline have been deliberately kept general, which means that in many cases relatively extensive calculations only have a marginal effect on the result. The verification concept could be simplified at a later date for specific applications in connection with an approval for a specific construction product.

## 7.2 Properties of CF sheets relevant to design

Numerous experimental studies involving concrete cylinders wrapped with CF sheets have revealed that the tensile strength ascertained in tests on strips of material are not achieved on the member. The ultimate strain $\varepsilon_{\mathrm{Lu}}$ in tensile tests on commercially available CF sheets is about 14–16 mm/m. Much lower values in the region of approx. 2–4 mm/m are recommended in the relevant design codes for the ultimate tensile stress, or rather the corresponding ultimate strain. Such figures should be used unless more accurate values are available. There are several reasons why the strain figure that can be used in confining reinforcement applications is much lower. One of these is the transverse compression acting on the fibre-reinforced material perpendicular to the direction of the fibres, which results from the longitudinal and transverse deformation of the loaded confined concrete member and the rounding radius $R_{\mathrm{c}}$.

Several writers have reported that a breakdown in the concrete microstructure of wrapped compression members leads to the formation of sharp-edged pieces that eventually cause local failures of the wrapping material (see also section 3.4.2). The specific behaviour of the confined concrete should therefore be seen as an additional influencing factor. It is also known that the relative ultimate strain for a wrapped reinforced concrete member is much lower than that of an unreinforced test specimen. The reason for this is the additional transverse compression that is transferred from the highly stressed longitudinal reinforcing bars to the confining reinforcement. Between the supports provided by the links or helical reinforcement, local outward buckling of the yielding longitudinal reinforcing bars is also prevented by the confining reinforcement. This problem has been investigated experimentally and analytically by *Tastani et al.* [128]. In the DAfStb guideline [1, 2] this influence is only taken into account empirically by reducing the ultimate strain, as shown in Figure 7.2.

Beyond a certain rounding radius $R_{\mathrm{c}}$, no influence on the related ultimate strain has been observed in either plain or reinforced concrete members, and this is taken into account in the design model of the DAfStb guideline [1, 2] by the factor $\alpha_{\mathrm{r}}$. If this limit value for the rounding radius is taken as 60 mm in all cases, then for column diameters relevant in practice, $D \geq 120$ mm, a mean related ultimate strain amounting to 0.5 and a characteristic value of 0.25 can be determined on the basis of numerous tests reported in the literature. This latter value is taken into account in the DAfStb guideline [1, 2] by the recommended system coefficient $[k_2]$.

When applied in several layers, the CF sheets are attached either in the form of single-ply rings or a multi-ply winding. Confining reinforcement in the form of a CF sheet is

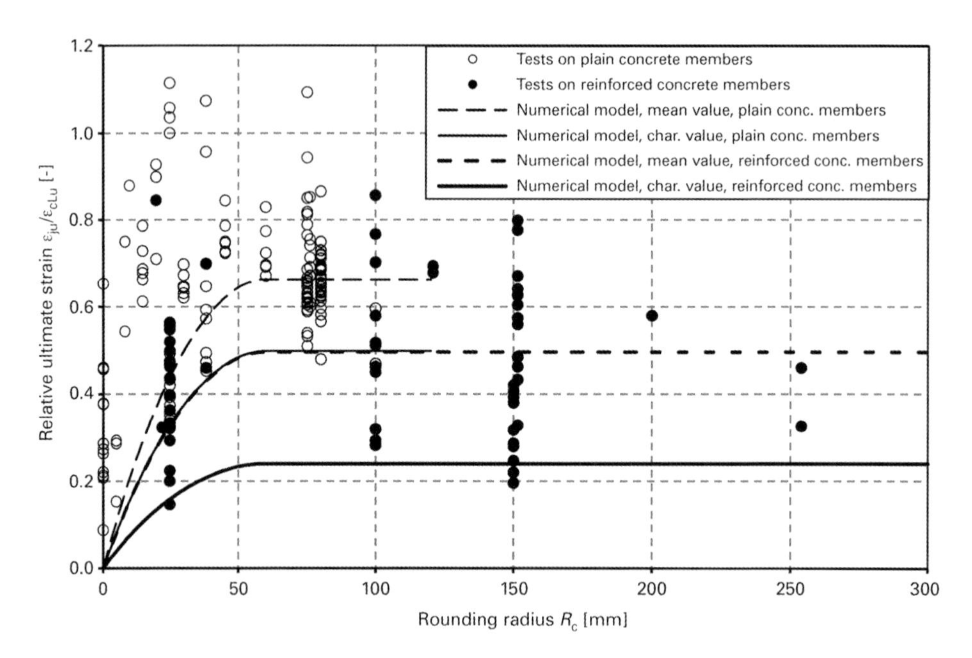


**Fig. 7.2** Ultimate strain $\varepsilon_{ju}$ in the CF sheet depending on the rounding radius $R_c$ for reinforced and plain concrete compression members

always finished with a lap. Owing to the disintegration of the concrete microstructure caused by the large deformations in the concrete, a loss of bond with the concrete surface must be assumed in the region of the lap. This means that the lap, without the help of the concrete, must be able to transfer the tensile force acting on one layer of the CF sheet. Therefore, the long-term and environmental influences that reduce the load-carrying capacity must be considered in the design. In the DAfStb guideline [1, 2] this is achieved by multiplying together the reduction factors for temperature $\alpha_T$, moisture $\alpha_F$, fatigue $\alpha_E$ and permanent loading $\alpha_Z$ as proposed by *Franke* and *Deckelmann* [108] for cold-curing epoxy resin adhesives. The value $[k_3]=0.7$ recommended for $\alpha_T$ in the DAfStb guideline [1, 2] was taken from the work of *Franke* and *Deckelmann* [108]. In the case of conventional applications in buildings, moisture and fatigue stresses can be ruled out, which means the system coefficients $[k_4]$ and $[k_5]$ can be recommended for the factors $\alpha_E$ and $\alpha_F$ respectively, taken as 1.0 in each case. A time of about 34 h, which is adequate for the ultimate limit state, has been verified experimentally for the ratio $\alpha_Z$ equal to the system coefficient $[k_6]=0.75$, which agrees with the results of the tests on epoxy resin concretes carried out by *Rehm et al.* [107]. This can also be seen in Figure 7.3, which compares the scatter given by *Rehm et al.* [107] with tests [129] carried out at the Technische Universität München.

Owing to the – in some circumstances – much higher level of stress compared with the un-strengthened condition, the strain in the CF sheet resulting from the creep-induced longitudinal deformation $\varepsilon_{cc}$ of the reinforced concrete column, which in turn can only be partially dissipated by the creep of the laminating resin, is also considered

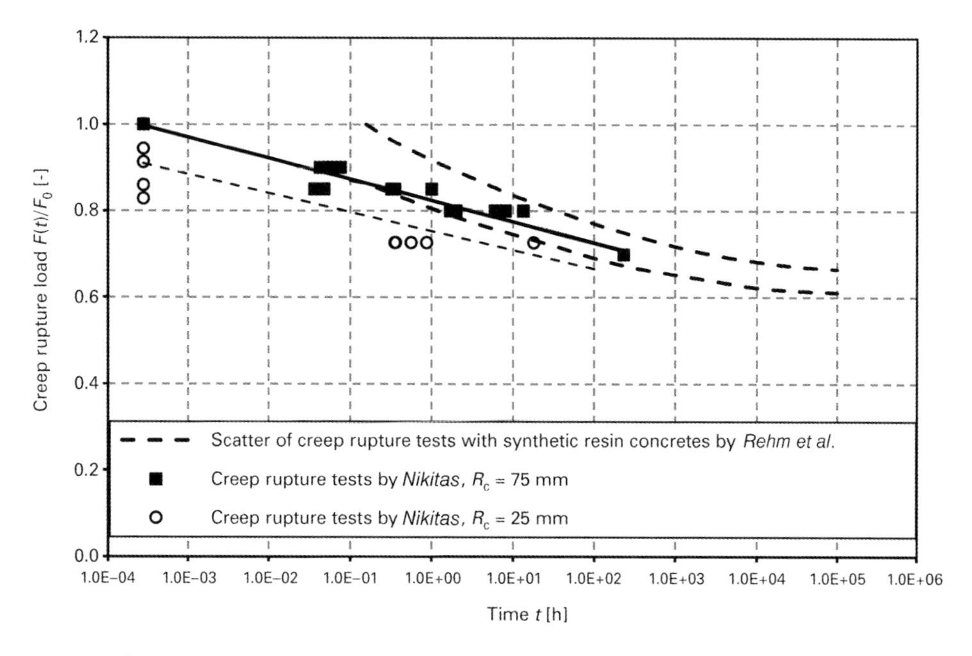


**Fig. 7.3** Creep rupture strength of CF sheets at laps

as an intrinsic stress state that reduces the load-carrying capacity. To consider this, the strain in the CF sheet is determined from the creep-induced longitudinal deformation $\varepsilon_{cc}$ in simplified form for a constant Poisson's ratio. The creep behaviour of a reinforced concrete column with a wrapping of CF sheet is investigated in more detail below. DIN EN 1992-1-1 [20, 21] specifies a Poisson's ratio $\nu = 0.2$ for uncracked concrete and according to the studies of *Lanig* [130] this represents an approximation that lies on the safe side for the stresses in strengthened reinforced concrete columns in the range of applications covered by the DAfStb guideline [1, 2]. The equation for determining the ultimate strain that may be assumed for a CF sheet is therefore

$$\varepsilon_{ju} = \alpha_r \cdot \alpha_T \cdot \alpha_F \cdot \alpha_E \cdot \alpha_Z \cdot \varepsilon_{Lu} - \nu \cdot \varepsilon_{cc} \tag{7.1}$$

## 7.3 Load-carrying capacity of cross-section

Many different approaches are possible when it comes to designing reinforced concrete compression members with a wrapping of fibre-reinforced polymer material. Some of those approaches have already been incorporated in national standards for the subsequent strengthening of reinforced concrete cross-sections, e.g. [119, 123]. In many instances the background to the approach is one of the models known from the fundamental work on confining compression members with reinforcing steel. However, as Figure 7.1 shows, these approaches are not entirely suitable for designing strengthening with fibre-reinforced polymers and need some form of modification at least.

In principle, we must distinguish between the approaches that are often based on simple empirical equations and require only manual calculations, and the much more complicated approaches that take into account equilibrium and compatibility conditions and are formulated with a view to being processed with the help of computer programs. In line with the objective of the DAfStb guideline [1, 2] to provide practical design methods, the following empirical addition function describes the relationship between the increase in the admissible compressive stress and the transverse compression $\sigma_l$ in a simple form. To this end, a factor $k_l = 2.0$ was derived from about 100 tests in order to determine the characteristic compressive strength of the confined concrete $f_{cck}$. The range of uniaxial concrete compressive strengths $f_{cm} \leq 58$ N/mm$^2$ investigated for this is used as the range of applicability for the method in the DAfStb guideline [1, 2].

$$f_{cck} = f_{ck} + k_l \cdot \sigma_l \tag{7.2}$$

The associated ultimate strain in the confined concrete can be determined using the following expression depending on the strain $\varepsilon_{c2}$ upon reaching the maximum strength of the concrete under uniaxial loading according to DIN EN 1992-1-1 [20, 21] and the transverse compressive stress $\sigma_l$ related to the mean value of the uniaxial concrete compressive strength $f_{cm}$:

$$\varepsilon_{cu} = \varepsilon_{c2} \cdot \left(1.75 + 19 \cdot \frac{\sigma_l}{f_{cm}}\right) \tag{7.3}$$

Reinforced concrete columns with a circular cross-section contain either links (ring-type ties) or helical steel reinforcement. If such columns are provided with additional confining reinforcement in the form of a CF sheet, the magnitude of the transverse compressive stress $\sigma_l$ is determined by the confining effect of both types of reinforcement. However, the respective range of influence of each confining reinforcement is different. The wrapping of fibre-reinforced material is positioned on the surface of the member, the reinforcing steel, on the other hand, confines only the core of the column within the helical or link reinforcing bars positioned with a certain concrete cover below the surface. Figure 7.4 shows the relationships within the cross-section for a compression member with both types of confining reinforcement.

We obtain the transverse compressive stress $p_1$ from the tensile force $F_L$ in the fibre-reinforced material. The higher transverse compressive stress $p_2$ due to the two forms of confining reinforcement acts within the reinforcing steel, which is considered to be smeared. To satisfy compatibility, the transverse compressive stress at the boundary between the concrete cover and the core of diameter $D_c$ cannot change abruptly. If as an approximation we presume a linear increase in the transverse compressive stress, then assuming that the transverse compressive stress $p_1$ results from the confining effect of the fibre-reinforced material only, the magnitude of the transverse compressive stresses can be determined as follows:

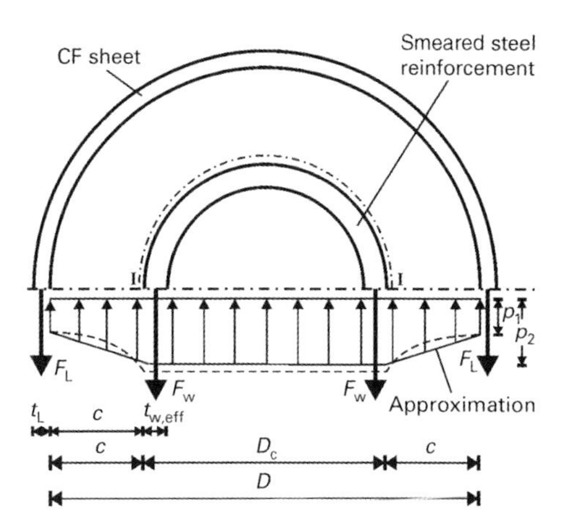


**Fig. 7.4** Transverse compression stresses in a confined compression member

$$p_1 = \frac{2 \cdot F_L}{D} = \frac{2 \cdot t_L \cdot E_L \cdot \varepsilon_{ju}}{D} \tag{7.4}$$

$$p_2 = \frac{2 \cdot (F_L + F_w) - p_1 \cdot c}{D_c + c} = \frac{2 \cdot \left(t_L \cdot E_L \cdot \varepsilon_{ju} + t_{w,eff} \cdot f_{wy}\right) - p_1 \cdot c}{D_c + c} \tag{7.5}$$

where:

$t_L$ theoretical thickness of fibre cross-section in CF sheet
$E_L$ modulus of elasticity of surface-mounted CF sheet relative to fibre cross-section
$\varepsilon_{ju}$ ultimate strain in fibre-reinforced material around member
$t_{w,eff}$ thickness of distributed confining reinforcing steel
$f_{wy}$ yield strength of confining reinforcing steel
$c$ concrete cover
$D$ diameter of reinforced concrete column
$D_c$ diameter of core area of column confined by reinforcing steel.

The decrease in transverse compression $\Delta p$ can be determined by considering a section I-I along the distributed reinforcing steel. The following applies:

$$2 \cdot t_L \cdot E_L \cdot \varepsilon_{ju} = (p_1 + p_2) \cdot c + \int_0^{\pi} (p_1 - \Delta p) \cdot \frac{D_c}{2} \cdot \sin \varphi \cdot d\varphi \tag{7.6}$$

$$\Delta p = p_1 - \frac{2 \cdot t_L \cdot E_L \cdot \varepsilon_{ju} - (p_1 + p_2) \cdot c}{D_c} \tag{7.7}$$

Concrete members in compression strengthened with fibre-reinforced materials exhibit varying behaviour depending on the intensity of the confining effect. The fundamental stress–strain curves shown in Figure 7.5 have been observed in experimental studies. Curve (0) describes the behaviour of an unconfined concrete compression member subjected to a uniaxial load in a short-term test with deformation control. Curve (1), for confined concrete, exhibits an only marginal increase in the maximum load. In a test

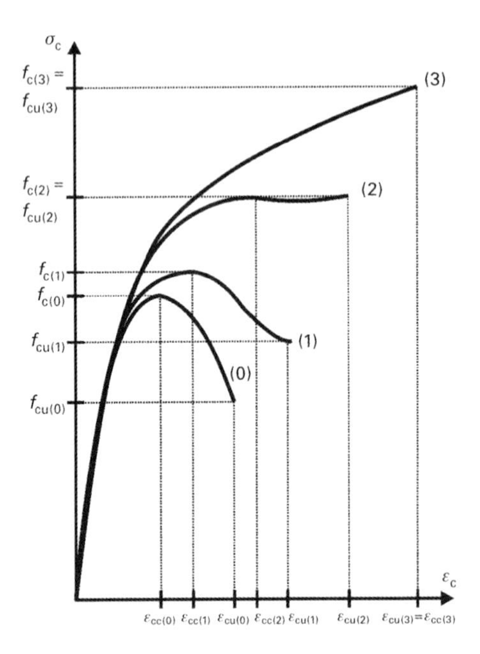


**Fig. 7.5** Potential stress–strain curves for compression members confined by fibre-reinforced materials and carrying axial loads

with deformation control, after the strain $\varepsilon_{cc(1)}$ associated with the maximum concrete compressive stress $f_{cc(1)}$ has been exceeded, the stress–strain curve descends until the confining reinforcement fails at strain $\varepsilon_{cu(1)}$, or rather the ultimate strength $f_{cu(1)}$ associated with that strain. Whereas curve (2) continues approximately horizontally after reaching the maximum concrete compressive stress $f_{cc(2)}$, in the case of curve (3) the strain and stress continue to rise until the confining reinforcement fails. Only in this latter case, where the stress–strain curve rises strictly monotonically, can we speak of an adequate confining effect with respect to activating the multi-axial strength of the concrete.

Various researchers have proposed criteria to guarantee an adequate confining effect. The experimental studies of *Xiao* and *Wu* [131] led them to propose a stiffness-related boundary between stress–strain curves rising strictly monotonically and stress–strain curves with a descending branch. According to [131], the descending branch does not occur when the following applies:

$$\frac{2 \cdot t_L \cdot E_L}{D \cdot f_c^2} \geq 0.2 \tag{7.8}$$

This approach has been incorporated in the DAfStb guideline.

Based on the work of *Eid* and *Paultre* [132], a simplified stress–strain curve was specified for design which, as proposed by *Lam* and *Teng* [133], is composed of a parabolic and a straight part and is continuously differentiable (see Figure 7.6). At the origin, the slope of the curve is given by the tangent modulus $E_c$ of the unconfined concrete. The design approach is defined by the following equations:

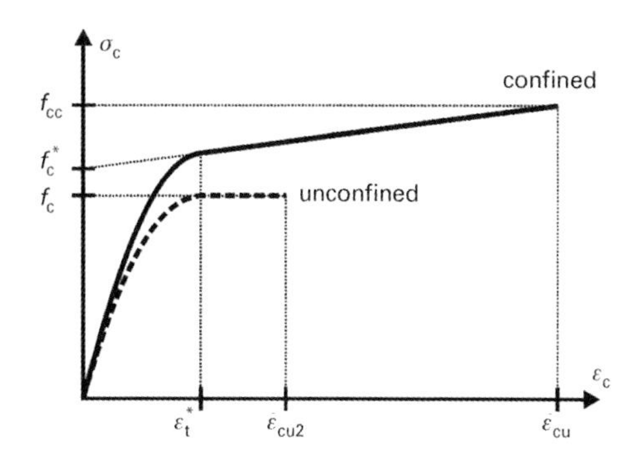


**Fig. 7.6** Simplified stress–strain curve for design

$$\sigma_c = \begin{cases} E_c \cdot \varepsilon_c - \dfrac{(E_c - E_2)^2}{4 \cdot f_c^*} \cdot \varepsilon_c^2 & \text{for } 0 \leq \varepsilon_c \leq \varepsilon_t^* \\ f_c^* + E_2 \cdot \varepsilon_c & \text{for } \varepsilon_t^* \leq \varepsilon_c \leq \varepsilon_{cu} \end{cases} \tag{7.9}$$

where:

$E_2$ slope of straight line according to Equation 7.10
$E_c$ tangent modulus of unconfined concrete subjected to uniaxial compression
$f_c^*$ point at which projected straight part of curve intersects stress axis according to Equation 7.11
$\sigma_c$ compressive stress in confined concrete
$\varepsilon_c$ longitudinal compressive strain in confined concrete
$\varepsilon_{cu}$ longitudinal strain in confined concrete at failure of fibre-reinforced material according to Equation 7.12.

$$E_2 = \frac{f_{cc} - f_c^*}{\varepsilon_{cu}} \tag{7.10}$$

$$f_c^* = f_c + [k_1] \cdot [\rho_{wy} \cdot f_{wy} - \Delta p] \cdot \left( \frac{D_c - \dfrac{s_w}{2}}{D} \right)^2 \tag{7.11}$$

$$\varepsilon_{cu} = \varepsilon_{c2} \cdot \left( 1.75 + 19 \cdot \frac{E_{jl} \cdot \varepsilon_{ju}}{f_{cm}} \right) \tag{7.12}$$

where:

$D$ diameter of reinforced concrete column
$D_c$ diameter of core area of column confined by reinforcing steel
$E_{jl}$ relative stiffness of confining reinforcement made from CF sheet
$f_{cc}$ admissible compressive stress in confined concrete at failure of fibre-reinforced material according to Equation 7.13
$f_{wy}$ yield strength of confining reinforcing steel

$s_w$ spacing of links or pitch of helical reinforcement in longitudinal direction of member
$\Delta p$ decrease in transverse compression as a result of different areas of influence of confining reinforcement according to Equation 7.7
$\rho_{wy}$ transverse reinforcing steel ratio according to Equation 7.14
$\varepsilon_{ju}$ assumed ultimate strain in CF sheet used as confining reinforcement around member.

$$f_{cc} = f_c + [k_1] \cdot \left[ E_{jl} \cdot \varepsilon_{ju} + \left( \rho_{wy} \cdot f_{wy} - \Delta p \right) \cdot \left( \frac{D_c - \frac{s_w}{2}}{D} \right)^2 \right] \quad (7.13)$$

$$\rho_{wy} = \frac{2 \cdot t_{w,eff}}{D_c} \quad (7.14)$$

$$E_{jl} = \frac{2 \cdot E_L \cdot t_L}{D} \quad (7.15)$$

where:

$E_L$ modulus of elasticity of surface-mounted CF sheet relative to fibre cross-section
$t_L$ theoretical thickness of fibre cross-section in CF sheet
$t_{w,eff}$ thickness of smeared confining reinforcing steel according to Equation 7.16.

$$t_{w,eff} = \frac{A_{sw}}{2 \cdot s_w} \quad (7.16)$$

where:

$A_{sw}$ total bar cross-section of effective confining transverse reinforcement per link or one complete winding of helical reinforcement.

The above equations for compressive strengths $f_{cc}$ and $f_c$* take into account the various areas of influence of the confinement in the form of CF sheet and reinforcing steel in a practical way. To do this, the concrete compressive stresses acting in the effective confined area within the confining reinforcing steel are distributed over the entire cross-section. At the same time, the effects of the individual steel links or helical reinforcement at a certain spacing/pitch in the longitudinal direction of the compression member are also taken into account through the theoretical notion of the parabolic arc according to *Sheikh* and *Uzumeri* [125].

However, only the effect of the confining CF sheet is used when defining the longitudinal strain $\varepsilon_{cc}$ in the confined concrete upon failure of the fibre-reinforced material.

## 7.4 Load-Carrying Capacity of Member

Most of the experimental studies of the load-carrying capacity of compression members with a wrapping of CF sheet were carried out on concrete cylinders with a height-to-diameter ratio of about 2 : 1. With fixity at both ends, which must be assumed for the majority of the tests, this corresponds to a slenderness ratio $\lambda = 4$. Therefore, the design approaches for the load-carrying capacity derived from these tests are only valid for members with similar geometrical conditions, i.e. small slenderness ratios. However, considerably greater slenderness ratios are found in practice; ratios between 20 and

35 are typical in buildings. For members regarded as slender, i.e. whose slenderness exceeds a certain value specified in the relevant design code, it should be realised that the load-carrying capacity of the member is not the same as that of the cross-section, but is in fact lower (see [134], for example).

The rules for strengthening reinforced concrete compression members in the relevant design codes either ignore the differences between the load-carrying capacity of the cross-section and the behaviour of the member, e.g. Concrete Society Technical Report No. 55 [120], or per definition are only valid for non-slender columns but do not provide any explicit definition of the maximum slenderness, e.g. ACI 440.R2-08 [119]. Further, as the load–deformation behaviour of the confined concrete is very different from the behaviour of conventional reinforced concrete columns, it can be assumed that the slenderness limits prescribed in the relevant standards for conventional reinforced concrete columns cannot be transferred to compression members wrapped with fibre-reinforced materials.

The Ph.D. thesis of *Jiang* [135] is a detailed treatment of the design of slender circular columns with a wrapping of fibre-reinforced material. In terms of the material behaviour of the confined concrete, *Jiang* assumes the curve proposed by *Lam* and *Teng* [133] for the confined concrete, which essentially corresponds to the simplified stress–strain curve in Figure 7.6 but ignores the confining effect of the reinforcing steel in the form of helical reinforcement or links. For the cross-section calculations, *Jiang* makes use of approaches for designing circular cross-sections subjected to axial forces and bending, which are based on an idealized stress distribution according to the stress block model and consider the smeared longitudinal reinforcement. The approaches formulated by *Jiang* include – differing from an exact stress calculation for the cross-section – a number of practical approximations that considerably simplify the calculation of the internal forces. *Jiang* combines the simplified approaches with the method for determining the deformation from the curvature of the member according to second-order theory, which is used in a similar way to, for example, the method with nominal curvature according to DIN EN 1992-1-1 [20, 21]. In *Jiang's* method the axial load $N_{\text{bal}}$ associated with the moment at maximum curvature $\phi_{\text{bal}}$ is calculated using the following expression, which was specified by *Jiang* empirically on the basis of a parametric study specifically for compression members with a wrapping of fibre-reinforced material. In contrast to unconfined compression members with a doubly symmetric cross-section, the maximum moment capacity is not reached at $N_{\text{bal}}$, but instead at lower axial loads.

$$N_{\text{bal}} = 0.8 \cdot f_{\text{cc}} \cdot \text{A} \tag{7.17}$$

where:

$f_{\text{cc}}$ compressive strength of confined concrete
$A$ gross cross-sectional area.

The following expression for the curvature $\phi_{\text{bal}}$ is valid for cross-sections with a rotationally symmetric arrangement of reinforcing steel:

$$\phi_{\text{bal}} = 2 \cdot \frac{\varepsilon_{\text{cu}} - \varepsilon_{\text{y}}}{D + D_c - (2 \cdot \phi_{\text{w}} + \phi_{\text{s}})} \tag{7.18}$$

where:

$\varepsilon_{cu}$ ultimate strain in confined concrete
$\varepsilon_y$ yield strain of longitudinal reinforcing steel: $\varepsilon_y = f_y/E_s$
$f_y$ yield strength of longitudinal reinforcing steel
$E_s$ modulus of elasticity of longitudinal reinforcing steel
$D$ diameter of reinforced concrete column
$D_c$ core diameter according to Figure 4.4
$\phi_w$ bar diameter of helical reinforcement or links
$\phi_s$ bar diameter of longitudinal reinforcing steel.

The equations for determining the theoretically admissible axial load $N_u$ and the associated moment $M_u$ according to *Jiang* [135] are as follows:

$$N_u = \theta \cdot \alpha_1 \cdot f_{cc} \cdot A_c \cdot \left(1 - \frac{\sin(2 \cdot \pi \cdot \theta)}{2 \cdot \pi \cdot \theta}\right) + (\theta_c - \theta_t) \cdot f_y \cdot A_s \quad (7.19)$$

$$M_u = N_u \cdot \left(e_1 + \left(\frac{l_0}{\pi}\right)^2 \cdot \xi_1 \cdot \xi_2 \cdot \phi_{bal}\right)$$
$$= \frac{2}{3} \cdot \alpha_1 \cdot f_{cc} \cdot A_c \cdot \frac{D}{2} \cdot \left(\frac{\sin^3(\pi \cdot \theta)}{\pi}\right) + f_y \cdot A_s \cdot \frac{D}{2} \cdot \frac{\sin(\pi \cdot \theta_c) + \sin(\pi \cdot \theta_t)}{\pi} \quad (7.20)$$

where:

$\theta$ relative angle describing the position of the stress block in the cross-section: $0 \leq \theta \leq 1$
$\alpha_1$ stress block geometry factor according to Equation 7.21
$f_{cc}$ compressive strength of confined concrete
$A_c$ gross cross-sectional area of concrete in reinforced concrete column
$\theta_c$ relative angle describing the stress distribution in the distributed longitudinal reinforcing steel subjected to compression: $0 \leq \theta_c = 1.25 \cdot \theta - 0.125 \leq 1$
$\theta_t$ relative angle describing the stress distribution in the distributed longitudinal reinforcing steel subjected to tension: $0 \leq \theta_t = 1.125 - 1.5 \cdot \theta \leq 1$
$f_y$ yield strength of longitudinal reinforcing steel
$A_s$ cross-sectional area of longitudinal reinforcing steel
$e_{tot}$ eccentricity of loading according to first-order theory: $e_{tot} = e_0 + e_i$
$e_0$ intentional eccentricity of loading according to first-order theory
$e_i$ additional unintentional eccentricity of loading according to the design codes
$\varepsilon_{ju}$ ultimate strain in CF sheet
$\varepsilon_{c2}$ longitudinal strain in concrete subjected to uniaxial compression upon reaching compressive strength
$D$ diameter of reinforced concrete column
$\phi_{bal}$ maximum curvature
$l_0$ buckling length of compression member
$\xi_1$ factor to allow for the decrease in curvature for a rise in the compressive force $N_u$ beyond $N_{bal}$ according to Equation 7.22

$\xi_2$ factor to allow for the geometry of the compression member and the strain in the confining reinforcement according to Equation 7.23.

$$\alpha_1 = 1.17 - 0.2 \cdot \frac{f_{cc}}{f_c} \tag{7.21}$$

$$\xi_1 = \frac{N_{bal}}{N_u} = \frac{0.8 \cdot f_{cc} \cdot A}{N_u} \leq 1 \tag{7.22}$$

$$\xi_2 = 1.15 + 0.06 \cdot \rho_\varepsilon - (0.01 + 0.012 \cdot \rho_\varepsilon) \cdot \frac{l_0}{D} \leq 1 \tag{7.23}$$

$$\rho_\varepsilon = \frac{\varepsilon_{ju}}{\varepsilon_{c2}} \tag{7.24}$$

where:

$f_c$ compressive strength of concrete subjected to uniaxial loading
$\rho_\varepsilon$ strain coefficient.

In order to determine the theoretically admissible axial load $N_u$, the relative angle $\theta$ must be calculated iteratively within the permissible range of answers by equating the two expressions (7.19) and (7.20) solved for $N_u$.

Figure 7.7 compares the theoretical load-carrying capacity calculated using the modified expressions (7.19) and (7.20) of *Jiang* with the results of the experimental studies of *Fitzwilliam* and *Bisby* [136] as well as *Ranger* and *Bisby* [137]. In their tests on

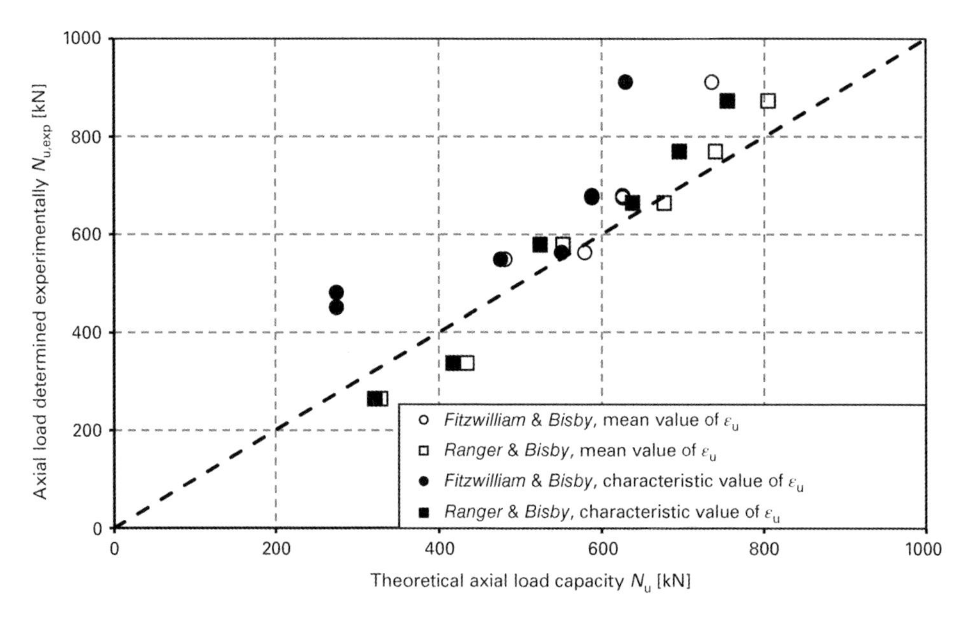


**Fig. 7.7** Comparison of the modified approach of Jiang for member loadbearing capacity with the results of the experimental studies by *Fitzwilliam* and *Bisby* [136] and *Ranger* and *Bisby* [137]

members, the researchers varied the slenderness, or rather the eccentricity of the axial load, over a range roughly coinciding with the applicability of the DAfStb guideline [1, 2]. The modification of *Jiang's* approach consisted of using the simplified stress–strain curve of Figure 7.6 for the design. Good agreement between model and tests has been observed; the influence of the individual parameters was also properly allowed for [56].

## 7.5 Creep

Strengthening measures that increase the load-carrying capacity of compression members by means of a wrapping of CF sheet do not increase the cross-section of the member. This inevitably leads to higher stresses in the concrete as a result of the larger actions. The work of *Rüsch* [138] over 50 years ago, likewise the later specific investigations of *Stöckl* [139] and other researchers, revealed a disproportionate increase in strain in connection with higher long-term loads exceeding about 40% of the uniaxial mean short-term compressive strength of the concrete. A linear relationship between the elastic deformation and the limit value for creep deformation $\varepsilon_{cc}(\infty, t_0)$ at time $t = \infty$ is generally assumed for lower creep-inducing stresses, which is expressed by the final creep coefficient $\varphi(\infty, t_0)$.

$$\varepsilon_{cc}(\infty, t_0) = \varphi(\infty, t_0) \cdot \frac{\sigma_c}{E_c} \tag{7.25}$$

where:

$\sigma_c$ creep-inducing longitudinal compressive stress
$E_c$ modulus of elasticity of concrete subjected to compression, which according to DIN EN 1992-1-1 [20, 21] is to be used as a tangent modulus.

With a view to avoiding disproportionate creep deformations, DIN EN 1992-1-1 [20, 21] specifies the following limit for concrete compressive stresses at the serviceability limit state:

$$\sigma_c \leq 0.45 \cdot f_{ck} \tag{7.26}$$

The disproportionate non-linear creep as a result of creep-inducing compressive stresses beyond this limit stress is described in DIN EN 1992-1-1 numerically using the following equation, which can be used to determine a modified final creep coefficient $\varphi_{nl}(\infty, t_0)$:

$$\varphi_{nl}(\infty, t_0) = \varphi(\infty, t_0) \cdot e^{\alpha_\sigma \cdot (k_\sigma - 0.45)} \tag{7.27}$$

where:

$\varphi(\infty, t_0)$ final creep coefficient for linear creep
$\alpha_\sigma$ stress intensity factor
$k_\sigma$ stress-strength ratio of concrete: $k_\sigma = \sigma_c / f_{ck}(t_0)$
$f_{ck}(t_0)$ characteristic concrete compressive stress at the time of loading.

DIN EN 1992-1-1 [20, 21] specifies a value of 1.5 for the stress intensity factor $\alpha_\sigma$.

With regard to the time-related behaviour of compression members wrapped with CF sheets, only the experimental studies of *Al Chami et al.* [140] were available at the time of carrying out the work on which the provisions of the DAfStb guideline [1, 2] are based. The provisions for describing the time- and load-dependent deformation behaviour of confined compression members in the DAfStb guideline [1, 2] were drafted on the basis of the relationships known from annex B of DIN EN 1992-1-1 [20, 21], taking into account the results of the experiments reported in [140]. A value of 2.7 was determined for the stress intensity factor $\alpha_\sigma$ on the basis of the results given in [140].

Generally, the creep coefficient $\varphi\,(t, t_0)$ according to duration of loading $(t, t_0)$ at the time considered $t$ is calculated as follows:

$$\varphi(t, t_0) = \varphi_0 \cdot \beta_c(t, t_0) \tag{7.28}$$

Here, the notional creep coefficient $\varphi_0$ is given by

$$\varphi_0 = \varphi_{RH} \cdot \beta(f_{cm}) \cdot \beta(t_0) \tag{7.29}$$

The coefficient $\varphi_{RH}$ describes the contribution of drying creep to the notional creep coefficient. For customary concrete members to DIN EN 1992-1-1 [20, 21], this coefficient is determined depending on the relative humidity of the ambient air, the geometry of the member and the compressive strength of the concrete. In their own series of tests, *Naguib* and *Mirmiran* [141] observed that a full wrapping with glass-fibre sheet saturated with epoxy resin functions like a vapour-tight sealing layer and therefore shrinkage deformations can be ignored. Drying creep, which makes a significant contribution to the creep deformation of unsealed concrete elements, is almost eliminated by a wrapping over the full surface area. The coefficient $\varphi_{RH}$ was therefore set to 1 for the full wrapping with CF sheets – applied using cold-curing, low-viscosity epoxy resins – dealt with in the DAfStb guideline [1, 2].

The coefficient $\beta(f_{cm})$ describes how the compressive strength of the concrete influences the notional creep coefficient depending on the mean cylinder compressive strength of the concrete $f_{cm}$ after 28 days. The formulation in EN 1992-1-1 [20] has been included in the DAfStb guideline [1, 2] without modification.

The system coefficient $[k_7]$ for $\beta\,(t_0)$ proposed in the DAfStb guideline [1, 2] in order to consider how the age of the concrete at the onset of loading influences the notional creep coefficient was specified as 0.39 – an average figure that results from the conditions of the experimental studies of *Al Chami et al.* [140]. The onset of loading in these tests was about four to six weeks after casting the specimens. Using this suggested value, the factors for the much higher concrete age and the preloading, which have a generally positive effect on strengthening measures, are ignored – an approach that lies on the safe side. However, suitable tests can be carried out on preloaded test specimens to establish a different, more realistic value. In such tests it may also be necessary to reproduce experimentally the spectrum of properties of the concrete aggregates relevant to creep for the range relevant in practice.

The following expression from Model Code 90 [142] was incorporated to describe the chronological development of creep after the onset of loading $\beta_c\ (t, t_0)$. In contrast to the corresponding relationship in DIN EN 1992-1-1 [20, 21], this formula also contains the time-based variable $t_1$. A reference time period $t_1 = 1.7$ d for describing the creep behaviour of concrete compression members wrapped with CF sheets was determined on the basis of the tests by *Al Chami et al.* [140].

$$\beta_c(t, t_0) = \left[\frac{(t - t_0)/t_1}{\beta_H + (t - t_0)/t_1}\right]^{0.3} \tag{7.30}$$

The coefficient $\beta_H$ takes into account the relative humidity of the air $RH$ and the effective member thickness relevant to drying creep $h_0$. Owing to the full wrapping, this coefficient is specified for the range of applicability of the DAfStb guideline [1, 2] without taking into account these two values and only depending on the compressive strength of the concrete.

$$\beta_H = \begin{cases} 250 & \text{for} f_{cm} \leq 35\ \text{N/mm}^2 \\ 250 \cdot \alpha_3 & \text{for} f_{cm} > 35\ \text{N/mm}^2 \end{cases} \tag{7.31}$$

$$\alpha_3 = \left[\frac{35}{f_{cm}}\right]^{0.5} \tag{7.32}$$

In the DAfStb guideline [1, 2] the creep deformation of reinforced concrete compression members with a circular cross-section and a full wrapping of CF sheet over the period of loading $\Delta t = t - t_0$ was related to the elastic deformation variable. The latter is determined from the creep-effective compressive stress $\sigma_{cp}$ and the modulus of elasticity, which in contrast to DIN EN 1992-1-1 [20, 21] is included as a secant modulus corresponding to the calibration of the model:

$$\varepsilon_{cc}(\Delta t) = \beta(t_0) \cdot \beta_c(\Delta t) \cdot \beta(f_{cm}) \cdot \beta_{0,k} \cdot \frac{\sigma_{cp}}{E_{cm}} \tag{7.33}$$

$$\beta_c(\Delta t) = \left[\frac{\Delta t/1.7}{\beta_H + \Delta t/1.7}\right]^{0.3} \tag{7.34}$$

$$\beta(f_{cm}) = \frac{16.8}{\sqrt{f_{cm}\left[\text{N/mm}^2\right]}} \tag{7.35}$$

$$\beta_{0k} = e^{2.7 \cdot (k_\sigma - 0.45)} \tag{7.36}$$

Figure 7.8 compares the approach of the DAfStb guideline [1, 2] with selected results from the work of *Al Chami et al.* [140]. This approach supplies an estimate that also lies on the safe side for the results of creep tests on reinforced concrete cylinders wrapped with aramid fibres by *Wang* and *Zhang* [143] and also the studies of *Berthet et al.* [144], which have in the meantime been published.

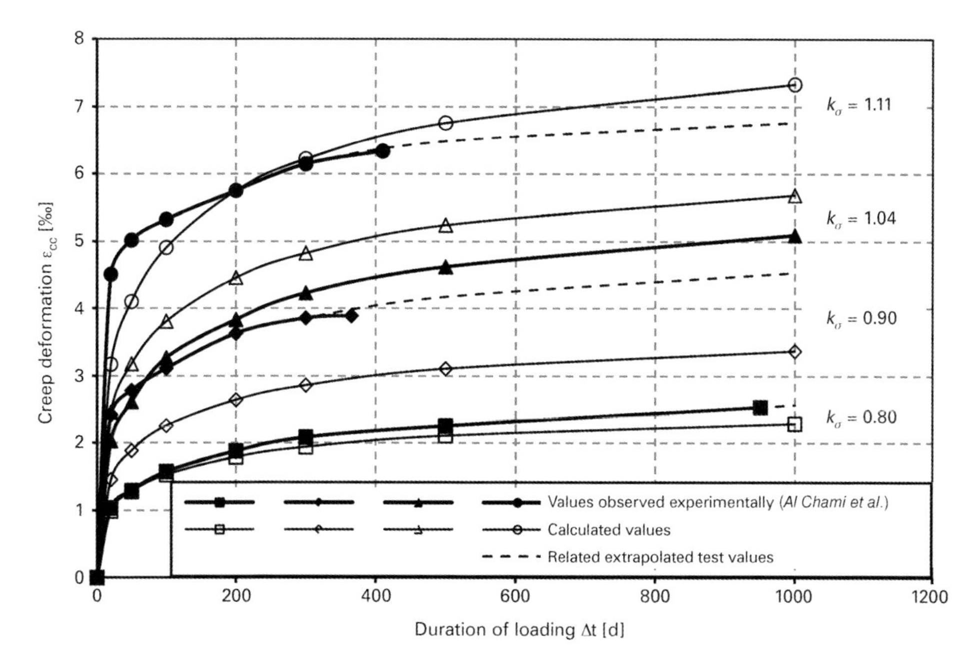


**Fig. 7.8** Comparison of development of creep deformations over time observed in the tests by *Al Chami et al.* [140] and calculated deformations

## 7.6 Analysis at Ultimate Limit State

The semi-empirical model of *Jiang* [135], extended by the expressions described above, will be used for the analysis at the ultimate limit state. To do this, the concrete compressive strength $f_{cc}$ of the reinforced concrete compression member with internal reinforcing steel and a wrapping of CF sheet will be used in conjunction with the equation for determining the ultimate strain that can be taken for the CF sheet. The effects of the creep of compression members with a wrapping of CF sheet are also taken into account in the ultimate limit state analyses required by the DAfStb guideline [1, 2]. Further, the effects of creep – also with respect to the increase in curvature – are considered in calculations according to second-order theory. This is taken into account practically according to section 5.8.3.3 of DIN EN 1992-1-1 [20, 21] by increasing the curvature by the factor $K_\varphi$, the value of which is linearly dependent on the effective final creep coefficient $\varphi_{ef}$. From the context of DIN EN 1992-1-1, this method can also be used in the range of non-linear creep, which is especially important for strengthened columns.

Of course, the characteristic values of the material strengths have been used in the formulations for design at the ultimate limit state. The creep deformation $\varepsilon_{cc}$ has a loadbearing-reduction effect on the strain that can be assumed for the confining reinforcement and so the factor $\alpha_k$ was introduced, which takes into account the coefficient of variation 0.3 commonly given in the literature (see [62], for example) for creep deformations. The partial safety factors given in section 2.4.2.4 of DIN EN

1992-1-1 or the DAfStb guideline [1, 2] are also used. As the failure of confined compression members within the scope of applicability of the DAfStb guideline [1, 2] can be attributed to the failure of the CF sheet, the expression for the load-carrying capacity of the confined concrete is reduced by the corresponding partial safety factor $\gamma_{LG}$.

The load-carrying capacity of a reinforced concrete column having a circular cross-section, a rotationally symmetrical arrangement of reinforcing steel and a wrapping of CF sheet can therefore be determined on the basis of the moment–axial load interaction, which is determined using the two equations below for calculating the resistances of the member. The first equation describes the axial load capacity $N_{Rd}$ and the second the associated moment capacity $M_{Rd}$ according to second-order theory.

$$N_{Rd} = \frac{1}{\gamma_{LG}} \cdot \theta \cdot \alpha_1 \cdot f_{cck} \cdot A_c \cdot \left(1 - \frac{\sin(2 \cdot \pi \cdot \theta)}{2 \cdot \pi \cdot \theta}\right) + \frac{1}{\gamma_s} \cdot (\theta_c - \theta_t) \cdot f_{yk} \cdot A_s \quad (7.37)$$

$$M_{Rd} = N_{Rd} \cdot \left(e_{tot} + \left(\frac{l_0}{\pi}\right)^2 \cdot \xi_1 \cdot \xi_2 \cdot \phi_{bal} \cdot K_\varphi\right)$$

$$= \frac{1}{\gamma_{LG}} \cdot \frac{2}{3} \cdot \alpha_1 \cdot f_{cck} \cdot A_c \cdot \frac{D}{2} \cdot \left(\frac{\sin^3(\pi \cdot \theta)}{\pi}\right)$$

$$+ \frac{1}{\gamma_s} \cdot f_{yk} \cdot A_s \cdot \frac{D}{2} \cdot \frac{\sin(\pi \cdot \theta_c) + \sin(\pi \cdot \theta_t)}{\pi} \quad (7.38)$$

where:

$\theta$ relative angle describing the position of the stress block in the cross-section: $0 \leq \theta \leq 1$

$\gamma_{LG}$ partial safety factor for surface-mounted CF sheet at the ultimate limit state

$\alpha_1$ stress block geometry factor.

$$\alpha_1 = 1.17 - 0.2 \cdot \frac{f_{cck}}{f_{ck}^*} \quad (7.39)$$

$f_{cck}$ characteristic compressive strength of confined concrete.

$$f_{cck} = f_{ck} + [k_1] \cdot \left[E_{jl} \cdot \varepsilon_{juk} + \left(\rho_{wy} \cdot f_{wyk} - \Delta p\right) \cdot \left(\frac{D_c - \frac{s_w}{2}}{D}\right)^2\right] \quad (7.40)$$

$f_{ck}$ compressive strength of concrete subjected to uniaxial loading

$E_{jl}$ relative stiffness of confining reinforcement made from CF sheet.

$$E_{jl} = \frac{2 \cdot E_L \cdot t_L}{D} \quad (7.41)$$

$E_L$ modulus of elasticity of surface-mounted CF sheet relative to fibre cross-section
$t_L$ theoretical thickness of fibre cross-section in CF sheet
$D$ diameter of reinforced concrete column
$\varepsilon_{juk}$ characteristic ultimate strain assumed for CF sheet.

$$\varepsilon_{juk} = [k_2] \cdot [k_3] \cdot [k_4] \cdot [k_5] \cdot [k_6] \cdot \varepsilon_{Lk} + \alpha_k \cdot \nu \cdot \varepsilon_{cc}(\Delta t) \tag{7.42}$$

$\varepsilon_{Lk}$ characteristic ultimate strain in CF sheet determined in tensile test on strip of CF material
$\alpha_k$ coefficient to allow for the increased scatter of creep deformations: $\alpha_k = 1.5$
$\nu$ Poisson's ratio: $\nu = 0.2$
$\varepsilon_{cc}(\Delta t)$ longitudinal deformation of reinforced concrete column due to creep.

$$\varepsilon_{cc}(\Delta t) = [k_7] \cdot \beta_c(\Delta t) \cdot \beta(f_{cm}) \cdot \beta_{0,k} \cdot \frac{\sigma_{cp}}{E_{cm}} \tag{7.43}$$

$\beta_c(\Delta t)$ coefficient for describing the development of creep over time.

$$\beta_c(\Delta t) = \begin{cases} 1 & \text{for normal strengthening tasks} \\ \left[\dfrac{\Delta t/1.7}{\beta_H + \Delta t/1.7}\right]^{0.3} & \text{members with short remaining lifetimes} \end{cases} \tag{7.44}$$

$\Delta t$ remaining lifetime [d]
$\beta_H$ coefficient for describing the influence of moisture.

$$\beta_H = \begin{cases} 250 & \text{for } f_{cm} \leq 35\ \text{N/mm}^2 \\ 250 \cdot \alpha_3 & \text{for } f_{cm} > 35\ \text{N/mm}^2 \end{cases} \tag{7.45}$$

$f_{cm}$ mean value of uniaxial concrete compressive strength [N/mm$^2$]
$\beta\,(f_{cm})$ coefficient to allow for the influence of the concrete compressive strength at the time of strengthening.

$$\beta(f_{cm}) = \frac{16.8}{\sqrt{f_{cm}}} \tag{7.46}$$

$\beta_{0,k}$ coefficient to allow for the loading level with respect to creep of the confined concrete.

$$\beta_{0,k} = \begin{cases} e^{(2.7 \cdot (k_\sigma - 0.45))} & \text{for } k_\sigma > 0.45 \\ 1 & \text{for } k_\sigma \leq 0.45 \end{cases} \tag{7.47}$$

$k_\sigma$ stress-strength ratio of concrete: $k_\sigma = \sigma_{cp}/f_{cm}$
$\sigma_{cp}$ creep-effective concrete compressive stress due to quasi-permanent actions.

$$\sigma_{cp} = \left|\frac{N_{Eqp}}{A_i}\right| + \left|\frac{M_{0Eqp}}{I_i}\right| \tag{7.48}$$

$N_{Eqp}$ axial load due to quasi-permanent actions at the serviceability limit state

$M_{0Eqp}$ moment according to first-order theory due to quasi-permanent actions at the serviceability limit state taking into account intentional and unintentional eccentricity

$A_i$ idealized cross-section of reinforced concrete column.

$$A_i = A_c + (\alpha_s - 1) \cdot A_s \tag{7.49}$$

$A_c$ gross cross-sectional area of concrete in reinforced concrete column

$\alpha_s$ modular ratio.

$$\alpha_s = \frac{E_s}{E_{cm}} \tag{7.50}$$

$E_s$ modulus of elasticity of longitudinal reinforcing steel

$E_{cm}$ secant modulus of concrete subjected to uniaxial compression at the time of strengthening

$I_i$ idealized moment of inertia of reinforced concrete column.

$$I_i = I_c + (\alpha_s - 1) \cdot \sum_j A_s^j \cdot \left(z_s^j\right)^2 \tag{7.51}$$

$I_c$ moment of inertia of gross concrete cross-section

$A_s^j$ cross-section of single reinforcing bar j

$z_s^j$ distance of single reinforcing bar j from centroid

$\rho_{wy}$ transverse reinforcing steel ratio.

$$\rho_{wy} = \frac{2 \cdot t_{w,eff}}{D_c} \tag{7.52}$$

$t_{w,eff}$ thickness of distributed transverse reinforcing steel.

$$t_{w,eff} = \frac{A_{sw}}{2 \cdot s_w} \tag{7.53}$$

$A_{sw}$ total bar cross-section of effective confining transverse reinforcement per link or one complete winding of helical reinforcement

$s_w$ spacing of links or pitch of helical reinforcement in longitudinal direction of member

$D_c$ diameter of core area of column confined by reinforcing steel

$f_{wyk}$ characteristic yield strength of transverse reinforcing steel

$\Delta p$ reduction in transverse compression due to the different areas of influence of the confining reinforcement.

$$\Delta p = p_1 - \frac{2 \cdot t_L \cdot E_L \cdot \varepsilon_{juk} - (p_1 + p_2) \cdot c}{D_c} \tag{7.54}$$

$c$ concrete cover

$p_1$ transverse compressive stress due to the confining effect of the CF sheet.

$$p_1 = \frac{2 \cdot t_L \cdot E_L \cdot \varepsilon_{juk}}{D} \tag{7.55}$$

$p_2$ transverse compressive stress due to the confining effect of the CF sheet and the transverse reinforcing steel.

$$p_2 = \frac{2 \cdot \left(t_L \cdot E_L \cdot \varepsilon_{juk} + t_{w,eff} \cdot f_{wyk}\right) - p_1 \cdot c}{D_c + c} \tag{7.56}$$

$f_{ck}^*$ point at which projected straight part of curve intersects stress axis.

$$f_{ck}^* = f_{ck} + k_1 \cdot \left[\rho_{wy} \cdot f_{wyk} - \Delta p\right] \cdot \left(\frac{D_c - \frac{s_w}{2}}{D}\right)^2 \tag{7.57}$$

$A_c$ gross cross-sectional area of concrete in reinforced concrete column
$\gamma_s$ partial safety factor for reinforcing steel at ultimate limit state
$\theta_c$ relative angle describing the stress distribution in the distributed longitudinal reinforcing steel subjected to compression: $0 \leq \theta_c = 1.25 \cdot \theta - 0.125 \leq 1$
$\theta_t$ relative angle describing the stress distribution in the distributed longitudinal reinforcing steel subjected to tension: $0 \leq \theta_t = 1.125 - 1.5 \cdot \theta \leq 1$
$f_{yk}$ characteristic yield strength of longitudinal reinforcing steel
$e_{tot}$ eccentricity of loading according to first-order theory: $e_{tot} = e_0 + e_i$
$e_0$ intentional eccentricity of loading according to first-order theory
$e_i$ additional unintentional eccentricity of loading to DIN EN 1992-1-1
$l_0$ buckling length of compression member
$\xi_1$ factor to allow for the decrease in curvature for a rise in the compressive force $N_{Rk}$ beyond $N_{bal}$.

$$\xi_1 = \frac{N_{bal}}{N_{Rk}} = \frac{0.8 \cdot f_{cck} \cdot A_c}{N_{Rd} \cdot \gamma_{LG}} \leq 1 \tag{7.58}$$

$\xi_2$ factor to allow for the geometry of the compression member and the strain in the confining reinforcement.

$$\xi_2 = 1.15 + 0.06 \cdot \rho_\varepsilon - (0.01 + 0.012 \cdot \rho_\varepsilon) \cdot \frac{l_0}{D} \leq 1 \tag{7.59}$$

$\rho_\varepsilon$ strain coefficient.

$$\rho_\varepsilon = \frac{\varepsilon_{juk}}{\varepsilon_{c2}} \tag{7.60}$$

$\varepsilon_{c2}$ longitudinal strain in concrete subjected to uniaxial compression upon reaching compressive strength: $\varepsilon_{c2} = 0.002$
$D$ diameter of reinforced concrete column
$\phi_{bal}$ maximum curvature.

$$\phi_{bal} = 2 \cdot \frac{\varepsilon_{cu} - \varepsilon_{yk}}{D + D_c - (2 \cdot \phi_w + \phi_s)} \tag{7.61}$$

$\varepsilon_{cu}$ longitudinal strain in confined concrete at failure of fibre-reinforced material.

$$\varepsilon_{cu} = \varepsilon_{c2} \cdot \left(1.75 + 19 \cdot \frac{E_{jl} \cdot \varepsilon_{juk}}{f_{cm}}\right) \tag{7.62}$$

$\varepsilon_{yk}$ strain in reinforcing steel upon reaching the characteristic yield strength: $\varepsilon_{yk} = f_{yk}/E_s$
$\phi_w$ bar diameter of transverse reinforcing steel
$\phi_s$ bar diameter of longitudinal reinforcing steel
$K_\varphi$ factor to allow for creep to DIN EN 1992-1-1.

$$K_\varphi = 1 + \beta \cdot \varphi_{ef} \tag{7.63}$$

$\beta$ factor to allow for the properties of the reinforced concrete column.

$$\beta = 0.35 + \frac{f_{ck}}{200} - \frac{\lambda}{150} \tag{7.64}$$

$\lambda$ slenderness of reinforced concrete column
$\varphi_{ef}$ effective creep coefficient.

$$\varphi_{ef} = [k_7] \cdot \beta(f_{cm}) \cdot \beta_{0,k} \cdot \frac{M_{0Eqp}}{M_{0Ed}} \tag{7.65}$$

$M_{0Ed}$ design value of acting bending moment according to first-order theory.

The system coefficients $[k_1]$ to $[k_9]$ and the values $E_L$, $t_L$, $\varepsilon_{Lk}$ must be taken from a national technical approval.

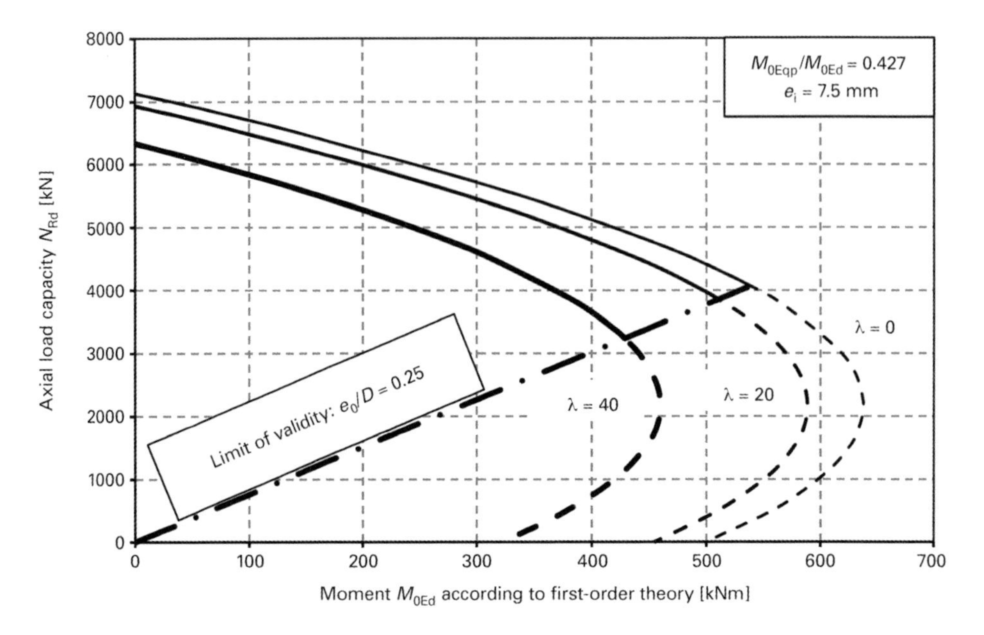


**Fig. 7.9** Influence of slenderness $\lambda$ on loadbearing capacity in M-N interaction diagram

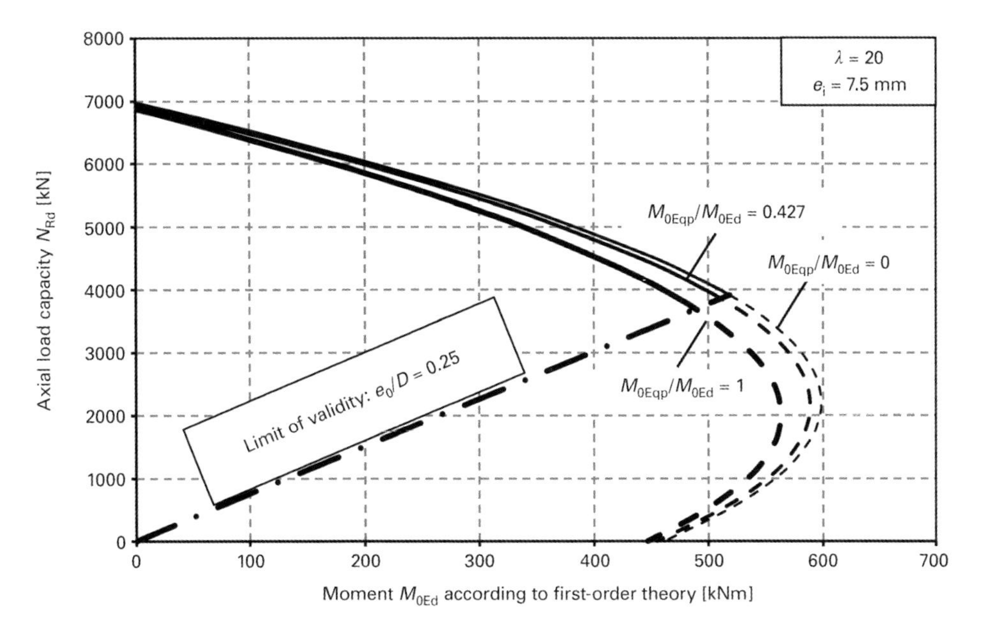


**Fig. 7.10** Influence of creep-relevant quasi-permanent loading component $M_{0Eqp}/M_{0Ed}$ on loadbearing capacity in M-N interaction diagram

The evaluation of the modified expressions of *Jiang* (7.19) and (7.20) enables M-N interaction diagrams to be drawn with relatively little effort. Figure 7.9 shows the decrease in a member's load-carrying capacity compared with the load-carrying capacity of its cross-section depending on slenderness, and Figure 7.10 the decrease compared with the creep-relevant quasi-permanent loading component $M_{0Eqp}/M_{0Ed}$ according to first-order theory. The empirical definition of the factor $\xi_1$ leads to a disproportionate increase in the curvature for axial loads $N_{Rk} < N_{bal}$ and, subsequently, to a distinct decrease in the theoretical flexural strength depending on the slenderness. This effect, which lies on the safe side, occurs with larger eccentricities and hence essentially lies outside the current scope of applicability. The confined column cross-section considered in the figures corresponds to the cross-section of the member in the design example considered in the next section.

## 7.7 Analysis at Serviceability Limit State

When it comes to serviceability, appropriate requirements regarding the design and detailing of strengthened reinforced concrete columns with a wrapping of fibre-reinforced materials are not known. The Technical Report No. 55 of the Concrete Society [120] limits the permissible compressive strain in the concrete to 3.5‰, mainly in order to avoid a brittle failure as a result of damage to already significantly stressed confining measures in the case of large longitudinal strains. As can be seen from Figure 7.5, the level of loading on the strengthened column for such a strain, which corresponds to the theoretical ultimate strain $\varepsilon_{cu2}$, is already significantly higher than the load-carrying capacity of the unstrengthened column, meaning that we must assume significant activation of the confining reinforcement. Therefore, this approach is unsuitable. On the other hand, ACI 440.2R-08 [119] contains recommendations for limiting the concrete stresses to 65% of the uniaxial compressive strength and the steel stresses to 60% of the yield strength. This level of stress is already reached with moderate concrete strengths and customary longitudinal reinforcement ratios under the rare load combination, which means many column strengthening measures would not comply with this criterion. It is also already known from the creep tests carried out on confined compression members that levels of loading that led to concrete stresses far in excess of the uniaxial compressive strength were able to be carried permanently over test periods of several years (see [140). Therefore, the provision in ACI 440.2R-08 [119] is not suitable either.

Verifying compliance with a maximum thickness for the confining reinforcement as required in the DAfStb guideline [1, 2] permits a level of loading on columns strengthened with a wrapping of CF sheet which is based on the load-carrying capacity of the unstrengthened reinforced concrete column. This procedure corresponds to the provisions of the earlier editions of DIN 1045 for compression members with confining helical reinforcement. The equation for the helical reinforcement ratio given in the 1988 edition of DIN 1045 [94] is attributed to *Müller* [112]:

$$A_w \cdot \beta_{Sw} \leq \delta \cdot \left[(2.3 \cdot A_b - 1.4 \cdot A_k) \cdot \beta_R + A_s \cdot \beta_S\right] \tag{7.66}$$

$$A_w = \frac{\pi \cdot d_k \cdot A_{sw}}{s_w} \tag{7.67}$$

$$A_k = \frac{\pi \cdot d_k^2}{4} \tag{7.68}$$

where:

$A_W$ cross-sectional area of helical steel reinforcement distributed over the column length
$d_k$ core diameter = diameter on centre-line of helical bar
$A_{sw}$ bar diameter of helix
$s_w$ pitch of helix
$\beta_{Sw}$ yield strength of helical reinforcement
$A_b$ total cross-section of compression member
$A_k$ core cross-section of compression member
$A_s$ total cross-section of longitudinal reinforcement
$\beta_R$ characteristic concrete compressive strength
$\beta_S$ steel stress at 2‰ compressive strain.

*Müller* defined the permissible level of stress in the confined compression member such that at the serviceability limit state (rare load combination), max. 80% of the calculative load-carrying capacity $N_u$ of the unconfined reinforced concrete cross-section without considering the ultimate strain could occur:

$$N_u = A_{bn} \cdot \beta_R + A_s \cdot \beta_S \tag{7.69}$$

where $A_{bn}$ is the net concrete cross-section.

This is effectively equal to limiting the structural response that can be assumed for the confining effect. The level of loading suggested by *Müller* [112] makes use of the provisions in the 1959 edition of DIN 1045 [145], which were intended to prevent the concrete shell outside the helical reinforcement from becoming detached as a result of cracking at the serviceability limit state. This rule was valid up until 2001 and can be regarded as proven, because no damage has been discovered to date.

The Equation 7.37 specified for the design value of the axial load capacity of a column wrapped with CF sheet was simplified for the required limit to the degree of strengthening.

$$\theta = 1.0 \tag{7.70}$$

$$\alpha_1 = 1.0 \tag{7.71}$$

$$1 - \frac{\sin(2 \cdot \pi \cdot \theta)}{2 \cdot \pi \cdot \theta} = 1.0 \tag{7.72}$$

$$\theta_c - \theta_t = 1.0 \tag{7.73}$$

Only the confining effect of the CF sheet is used for the compressive strength of the confined concrete:

$$f_{cck} = f_{ck} + [k_1] \cdot \left[ \frac{2 \cdot E_L \cdot t_L}{D} \cdot \varepsilon_{juk} \right] \tag{7.74}$$

The theoretical load-carrying capacity (characteristic value) of the unconfined reinforced concrete column to DIN EN 1992-1-1 [20], considering the strain compatibility, is

$$N_{Rk}^{0} = A_c \cdot \alpha_{cc} \cdot f_{ck} + A_s \cdot \varepsilon_{c2} \cdot E_s \tag{7.75}$$

where:

$A_c$ concrete cross-section
$\alpha_{cc}$ reduction factor for uniaxial concrete compressive strength in structure: $\alpha_{cc} = 0.85$
$f_{ck}$ characteristic uniaxial concrete compressive strength
$A_s$ total cross-section of longitudinal reinforcement
$\varepsilon_{c2}$ value of permissible compressive strain in concrete according to DIN EN 1992-1-1 [20] Tab. 3.1; taking into account the favourable effect of the creep of the concrete is permitted for small eccentricities according to DIN EN 1992-1-1/NA (NCI) [21] section 6.1 (3)P
$E_s$ modulus of elasticity of longitudinal reinforcing steel: $E_s = 200\ \text{kN/mm}^2$.

Comparing the characteristic value of the theoretical load-carrying capacity of the unconfined reinforced concrete column to DIN EN 1992-1-1 [20, 21] with the stipulation by *Müller* [112] leads to the following expression in the range of permissible longitudinal reinforcement ratios, which is only approximately linearly dependent on the compressive strength of the concrete:

$$\frac{0.8 \cdot N_{\mathrm{u}}}{N_{\mathrm{Rk}}^{0}} = [k_8] - [k_9] \cdot f_{\mathrm{ck}} \tag{7.76}$$

The recommended system coefficients $[k_8]$ and $[k_9]$ for the level of loading proposed by *Müller* [112] are

$$[k_8] = 0.89 \tag{7.77}$$

$$[k_9] = 0.0044 \tag{7.78}$$

Within the scope of a national technical approval, however, other values can be verified experimentally by way of suitable creep rupture tests.

With the above simplifications and the use of the partial safety factors $\gamma_{\mathrm{LG}}$ and $\gamma_{\mathrm{s}}$ on the resistance side plus $\gamma_{\mathrm{F}}$ on the actions side, we get the following as a condition for the load-carrying capacity of the confined column at the serviceability limit state:

$$([k_8] - [k_9] \cdot f_{\mathrm{ck}}) \cdot (A_{\mathrm{c}} \cdot \alpha_{\mathrm{cc}} \cdot f_{\mathrm{ck}} + A_{\mathrm{s}} \cdot \varepsilon_{\mathrm{c2}} \cdot E_{\mathrm{s}}) \geq \frac{1}{\gamma_{\mathrm{F}}} \cdot \left[\frac{1}{\gamma_{\mathrm{LG}}} \cdot \left(f_{\mathrm{ck}} + [k_1] \cdot \frac{2 \cdot E_{\mathrm{L}} \cdot t_{\mathrm{L}}}{D} \cdot \varepsilon_{\mathrm{juk}}\right) \cdot A_{\mathrm{c}} + \frac{1}{\gamma_{\mathrm{s}}} \cdot f_{\mathrm{yk}} \cdot A_{\mathrm{s}}\right] \tag{7.79}$$

Here, $\gamma_{\mathrm{F}}$ is the weighted partial safety factor for the actions corresponding to the contributions of the permanent and variable loads to the internal forces combination at the ultimate limit state. Solving for the theoretical thickness of the fibre cross-section of the confining reinforcement $t_{\mathrm{L}}$ results in the following expression given in the DAfStb guideline [1, 2]:

$$t_{\mathrm{L}} \leq \frac{D}{2 \cdot E_{\mathrm{L}} \cdot \varepsilon_{\mathrm{juk}}} \cdot \frac{1}{[k_1]} \cdot \left[\gamma_{\mathrm{LG}} \cdot \left[\gamma_{\mathrm{F}} \cdot ([k_8] - [k_9] \cdot f_{\mathrm{ck}}) \cdot \left(\alpha_{\mathrm{cc}} \cdot f_{\mathrm{ck}} + \frac{A_{\mathrm{s}}}{A_{\mathrm{c}}} \cdot |\varepsilon_{\mathrm{c2}}| \cdot E_{\mathrm{s}}\right) - \frac{f_{\mathrm{yk}}}{\gamma_{\mathrm{s}}} \cdot \frac{A_{\mathrm{s}}}{A_{\mathrm{c}}}\right] - f_{\mathrm{ck}}\right] \tag{7.80}$$

This criterion primarily governs when the longitudinal reinforcing steel only makes a minor contribution to the load-carrying capacity, i.e. in the case of low reinforcement ratios in combination with high uniaxial concrete compressive strengths. However, in the majority of the applications for which the DAfStb guideline [1, 2] is valid, provision RV 3.10.3 of the guideline governs. This provision is based on experience and specifies that a maximum of 10 layers of CF sheet material may be attached. Together, the two criteria guarantee that the loads on the concrete remain within the domain of experimental evidence. In addition, redistribution to the longitudinal reinforcing steel is possible, which leads to lower concrete stresses and a rapid reduction in the creep deformations.

# 8 Example 3: Column strengthening

## 8.1 System

### 8.1.1 General

Owing to a change of use, a reinforced concrete column in a residential building needs to be strengthened. As-built documents with structural calculations to DIN 1045 [146] are available. A wrapping of CF sheet is to be used for strengthening the column. Figure 8.1 shows the structural system requiring strengthening.

### 8.1.2 Loading

The loads are predominantly static. Three load cases will be investigated for ultimate limit state design:

- **Load case 1** represents the situation prior to strengthening.
- **Load case 2** is the loading during strengthening. The strengthening measures are carried out under the dead load of the member. Existing fitting-out items will be removed during the strengthening work.
- **Load case 3** represents the loading situation in the strengthened condition.

The actions for the various load cases are listed in Table 8.1. Furthermore, the eccentricity of loading due to imperfections must be considered according to DIN EN 1992-1-1 [20, 21] section 5.2 (7):

$$e_i = l_0/400 = 3000/400 = 7.5\,\text{mm}$$

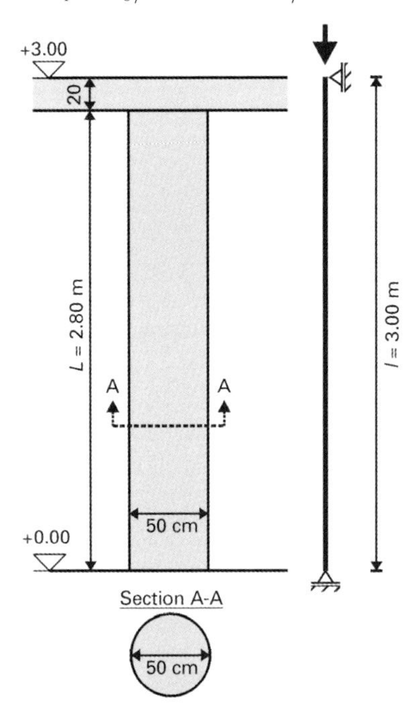


**Fig. 8.1** Column system requiring strengthening

**Table 8.1** Loads on the system in kN/m² for the various load cases.

| Load case | 1 | 2 | 3 |
|---|---|---|---|
| $G_k$ (dead load + fitting-out) | 2014.7 | 2014.7 | 2014.7 |
| $Q_k$ (imposed load, category B) | 1400 | — | 2500 |

Load case 3 governs for designing the strengthening measures. The load combination for the ultimate limit state and the load combination for the serviceability limit state under a quasi-permanent load combination are required for the analyses. These load combinations are given by DIN EN 1990 [24] together with its associated National Annex [25]. The load at the ultimate limit state (persistent and transient design situations) for this example is

$$\sum_{j\geq 1} \gamma_{G,j} \cdot G_{k,j} + \gamma_P \cdot P + \gamma_{Q,1} \cdot Q_{k,1} + \sum_{i>1} \gamma_{Q,i} \cdot \psi_{0,i} \cdot Q_{k,i}$$

$$N_{Ed} = \gamma_G \cdot G_k + \gamma_Q \cdot Q_k = 1.35 \cdot 2014.7 + 1.5 \cdot 2500 = 6469.8\,\text{kN}$$

and the loading for the serviceability limit state under a quasi-permanent load combination is

$$\sum_{j\geq 1} G_{k,j} + P + \sum_{i>1} \psi_{2,i} \cdot Q_{k,i}$$

$$N_{Eqp} = G_{k,j} + \psi_2 \cdot Q_k = 2014.7 + 0.3 \cdot 2500 = 2764.7\,\text{kN}$$

### 8.1.3 Construction materials

#### 8.1.3.1 Concrete

Concrete of strength class C30/37 was able to be ascertained from the as-built documents according to DIN 1045 [146]. Following a test on the member to [95], the result was also class C30/37. Therefore, the values according to DIN EN 1992-1-1 [20] Table 3.1 will be used in the design, i.e. mean concrete compressive strength $f_{cm} = 38\,\text{N/mm}^2$, characteristic concrete compressive strength $f_{ck} = 30\,\text{N/mm}^2$ and modulus of elasticity $E_{cm} = 33\,\text{kN/mm}^2$.

#### 8.1.3.2 Type and quantity of existing reinforcement

According to the as-built documents, the longitudinal reinforcement is 12 Ø25 mm ($A_s = 58.9\,\text{cm}^2$) and the links are Ø10 mm @ 30 cm (closer at the ends – see Figure 8.2, $a_{sw}/s_w = 5.23\,\text{cm}^2/\text{m}$). With as-built documents to DIN 1045-1 [146], we can assume that the existing reinforcing steel has a yield stress $f_{syk} = 500\,\text{N/mm}^2$ and a modulus of elasticity $E_s = 200\,\text{kN/mm}^2$. Figure 8.2 shows the layout of the existing reinforcement.

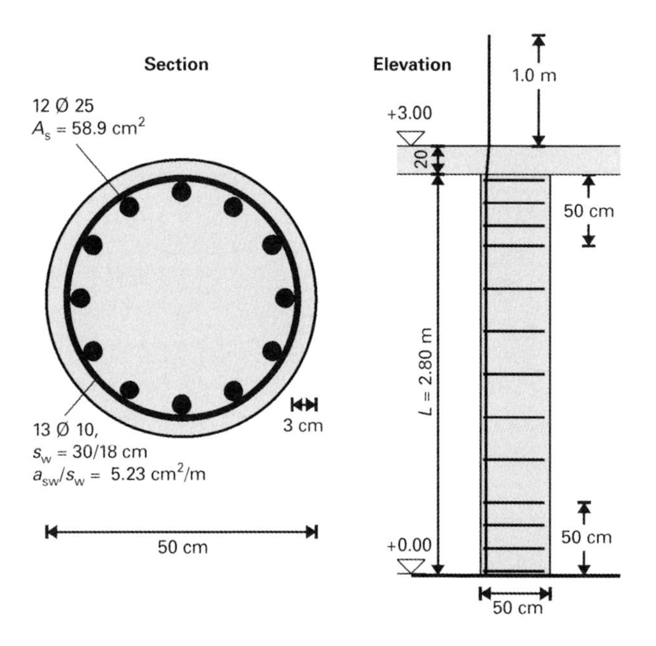


**Fig. 8.2** Type and position of existing reinforcement

#### 8.1.3.3 Strengthening system

Externally bonded CF sheet, tensile strength $f_{\mathrm{Luk}} = 3500\,\mathrm{N/mm^2}$ and modulus of elasticity $E_{\mathrm{L}} = 230\,000\,\mathrm{N/mm^2}$, will be used for the strengthening. This results in the following ultimate strain:

$$\varepsilon_{\mathrm{Lk}} = 3500/230 = 15.22\,\mathrm{mm/m}$$

One layer of CF sheet has a nominal thickness $t_{\mathrm{L,i}} = 0.15\,\mathrm{mm}$. The strengthening system includes an appropriate epoxy resin adhesive. According to DAfStb guideline part 1 section 2.1.1 (RV 4) [1, 2], the strengthening system must have a national technical approval. This approval defines the material properties but also further system coefficients, which for the system used here are listed in Table 8.2.

**Table 8.2** System coefficients.

| $[k_0]$ | $[k_1]$ | $[k_2]$ | $[k_3]$ | $[k_4]$ |
|---|---|---|---|---|
| $0.2 \cdot 1/(\mathrm{N/mm^2})$ | 2.0 | 0.25 | 0.7 | 1.0 |
| $[k_5]$ | $[k_6]$ | $[k_7]$ | $[k_8]$ | $[k_9]$ |
| 1.0 | 0.75 | 0.39 | 0.89 | $0.44 \cdot 10^{-2}$ |

**Table 8.3** Axial loads and bending moments for the relevant load combinations.

| Load combination | $N$ | $M$ |
|---|---|---|
| — | kN | kNm |
| Load case 3; ULS | 6469.8 | 48.5 |
| Load case 3; SLS, quasi-permanent | 2764.7 | 20.7 |

## 8.2 Internal forces

The above loads result in the axial loads on the column as given in Table 8.3, which also lists the moments due to axial load and eccentricity of loading.

## 8.3 Determining the cross-sectional values

The cross-sectional values of the column are required at several points in order to determine its load-carrying capacity. First of all, we need the modular ratio and the area of the concrete cross-section:

$$\alpha_s = \frac{E_s}{E_{cm}} = \frac{200}{33} = 6.1$$

$$A_c = D^2/4 \cdot \pi = 250^2 \cdot \pi = 1964 \cdot 10^2 \text{ mm}^2$$

Using these values it is possible to calculate the idealized area of the cross-section according to DAfStb guideline [1, 2] part 1, RV 6.1.4.2, Eq. (RV 6.85):

$$A_i = A_c + (\alpha_s - 1) \cdot A_s = 1964 \cdot 10^2 + (6.1 - 1) \cdot 58.9 \cdot 10^2 = 2264 \cdot 10^2 \text{ mm}^2$$

To calculate the idealized second moment of area, we first need the second moment of area of the gross concrete cross-section:

$$I_c = D^4/16 \cdot \frac{\pi}{4} = 250^4 \cdot \frac{\pi}{4} = 3068.0 \cdot 10^6 \text{ mm}^4$$

DAfStb guideline [1, 2] part 1, RV 6.1.4.2, Eq. (RV 6.87) is used to calculate the idealized second moment of area:

$$I_i = I_c + (\alpha_s - 1) \cdot \sum_j z_s^{j^2} \cdot A_s^j$$

As can be seen from the equation, $z_s^j$ and $A_s^j$ must be determined. This is carried out below according to Figure 8.3.

First of all we must determine the positions of the bars, or rather the radii to the centres of the reinforcing bars. This depends on the concrete cover and the diameter of the links $\phi_{sw}$ and the bars $\phi_s$.

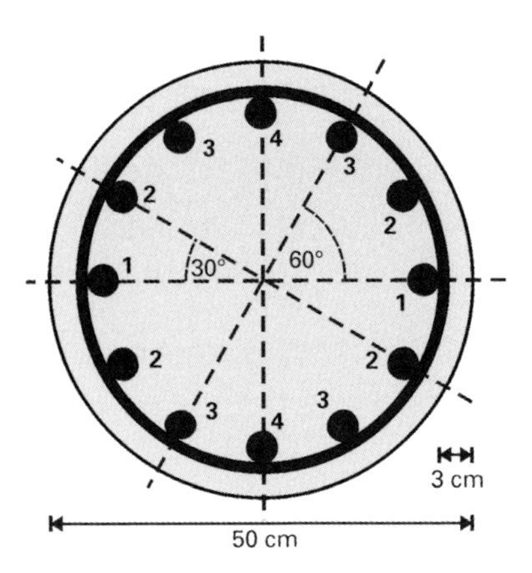


**Fig. 8.3** Scheme for determining $z_s^j$ and $A_s^j$

$$r_s = D/2 - c - \varphi\phi_{sw} - \frac{\phi_s}{2} = 250 - 30 - 10 - \frac{25}{2} = 197.5 \text{ mm}$$

Consequently, we get the following figures for the bars marked with number 1:

$$A_{s,1} = 2 \cdot \left(\frac{25}{2}\right)^2 \cdot \pi = 981.7 \text{ mm}^2 \quad z_{s,1} = \sin(0°) \cdot r_s = 0$$

We proceed similarly for the bars marked with numbers 2, 3 and 4:

$$A_{s,2} = 4 \cdot \left(\frac{25}{2}\right)^2 \cdot \pi = 1963.5 \text{ mm}^2 \quad z_{s,2} = \sin(30°) \cdot r_s = 0.5 \cdot 197.5 = 98.8 \text{ mm}$$

$$A_{s,3} = 4 \cdot \left(\frac{25}{2}\right)^2 \cdot \pi = 1963.5 \text{ mm}^2 \quad z_{s,3} = \sin(60°) \cdot r_s = 0.866 \cdot 197.5 = 171.0 \text{ mm}$$

$$A_{s,4} = 2 \cdot \left(\frac{25}{2}\right)^2 \cdot \pi = 981.7 \text{ mm}^2 \quad z_{s,4} = \sin(90°) \cdot r_s = 197.5 \text{ mm}$$

It is now possible to calculate the idealized second moment of area of the column cross-section:

$$I_i = 3068.0 \cdot 10^6 + (6.1 - 1) \cdot (98.8^2 \cdot 1963.5 + 171.0^2 \cdot 1963.5 + 197.5^2 \cdot 981.7)$$
$$= 3653.9 \cdot 10^6 \text{ mm}^4$$

## 8.4 Boundary conditions

Prior to calculating the load-carrying capacity of the column, it is first necessary to check whether the column may be strengthened using the method given in the guideline.

The minimum column diameter complies with DAfStb guideline [1, 2] part 1, RV 6.1.4.2, Eq. (RV 6.58):

$$D = 500\,\text{mm} \geq 120\,\text{mm}$$

Likewise, the permissible slenderness and permissible intentional eccentricity of loading comply with DAfStb guideline [1, 2] part 1, RV 6.1.4.2, Eqs. (RV 6.59) and (RV 6.60) respectively:

$$\lambda = \frac{l_0}{i} = \frac{1 \cdot l}{\sqrt{I_\text{i}/A_\text{i}}} = \frac{1 \cdot 3000}{\sqrt{3562.0 \cdot 10^6/226.3 \cdot 10^3}} = 23.9 \leq 40$$

$$\frac{e_0}{D} = \frac{0}{500} = 0 \leq 0.25$$

The permissible concrete compressive strength according to DAfStb guideline [1, 2] part 1, RV 6.1.4.2, Eq. (RV 6.61) is not exceeded either:

$$f_\text{cm} = 38\,\text{N/mm}^2 \leq 58\,\text{N/mm}^2$$

According to DAfStb guideline [1, 2] part 1, RV 6.1.4.2, Eq. (RV 6.57), a minimum thickness $t_\text{L}$ is necessary in order to activate the multi-axial stress state by way of a sufficient confining effect:

$$t_\text{L} \geq \frac{[\text{k}_0] \cdot D \cdot f_\text{cm}^2}{E_\text{L}} = \frac{0.2 \cdot 500 \cdot 38^2}{230\,000} = 0.63\,\text{mm}$$

A thickness $t_\text{L}$ of 0.63 mm leads to a number of layers $n_\text{L} = 0.63/0.15 = 4.2$; thus, the wrapping will consist of five layers. The column may therefore be strengthened in accordance with the method given in the DAfStb guideline.

## 8.5 Verification of column load-carrying capacity

A number of layers $n_\text{L} = 5$ and the following fabric thickness are used in the analyses given below:

$$t_\text{L} = n_\text{L} \cdot t_\text{L,i} = 5 \cdot 0.15 = 0.75\,\text{mm}$$

### 8.5.1 Creep of confined concrete

In order to consider the behaviour with respect to time, the creep of the confined concrete must be calculated according to DAfStb guideline [1, 2] part 1 Eq. (RV 6.74):

$$\varepsilon_\text{cc}(\Delta t) = [k_7] \cdot \beta_\text{c}(\Delta t) \cdot \beta(f_\text{cm}) \cdot \beta_\text{0,k} \cdot \frac{\sigma_\text{cp}}{E_\text{cm}}$$

One of the things needed for this is the compressive stress relevant to creep under a quasi-permanent load combination, which is calculated according to DAfStb guideline [1, 2] part 1 Eq. (RV 6.84):

$$\sigma_{cp} = \left|\frac{N_{0,Eqp}}{A_i}\right| + \left|\frac{M_{0,Eqp}}{I_i \cdot \frac{2}{D}}\right| = \left|\frac{2764.7 \cdot 10^3}{2264 \cdot 10^2}\right| + \left|\frac{20.7 \cdot 10^6}{3653.9 \cdot 10^6 \cdot \frac{2}{500}}\right| = 13.63\,\text{N/mm}^2$$

The factor $\beta_c(\Delta t)$ takes into account the chronological development of the action. As this is a customary strengthening assignment, we shall select $\beta_c(\Delta t) = 1.0$ according to DAfStb guideline [1, 2] part 1 Eq. (RV 6.75).

It is also important to check whether non-linear creep needs to be considered. To do this, the degree of utilization under a quasi-permanent load must be calculated. As this is $< 0.45$, it is not necessary to consider non-linear creep and therefore $\beta_{0,k} = 1$ (see DAfStb guideline [1, 2] part 1 Eqs. (RV 6.82) and (RV 6.83)).

$$k_\sigma = \frac{\sigma_{cp}}{f_{cm}} = \frac{13.63}{38} = 0.36 \leq 0.45$$

Likewise, the influence of the concrete compressive strength at the time of strengthening must be considered according to DAfStb guideline [1, 2] part 1 Eq. (RV 6.80):

$$\beta(\text{f}_{cm}) = \frac{16.8}{\sqrt{f_{cm}}} = \frac{16.8}{\sqrt{38}} = 2.73$$

With these factors and factor [k7] from Table 8.2 at our disposal, we can now calculate the creep of the confined concrete:

$$\varepsilon_{cc}(\Delta t) = 0.39 \cdot 1 \cdot 2.73 \cdot 1 \cdot \frac{13.63}{33\,000} = 0.44\,\text{mm/m}$$

### 8.5.2 Properties of the CF sheet

The stiffness of the wrapping relative to the column diameter $D$ is required for the next calculations. According to DAfStb guideline [1, 2] part 1 Eq. (RV 6.66), this relative stiffness is

$$E_{jl} = \frac{2 \cdot E_L \cdot t_L}{D} = \frac{2 \cdot 230\,000 \cdot 0.75}{500} = 690\,\text{N/mm}^2$$

We also need the long-term characteristic value of the strain in the confining reinforcement, which is calculated according to DAfStb guideline [1, 2] part 1 Eq. (RV 6.67). This is calculated depending on the coefficients for considering the influence of the change of direction at the member $\alpha_r = [k_2] = 0.25$, temperature $\alpha_T = [k_3] = 0.7$, moisture $\alpha_F = [k_4] = 1.0$, type of loading $\alpha_E = [k_5] = 1.0$, duration of loading $\alpha_Z = [k_6] = 1.0$ and the creep of the confined concrete $\varepsilon_{cc}(\Delta t)$ (Section 8.5.1), the factor for taking into account the scatters of the creep deformations $\alpha_k = 1.5$,

Poisson's ratio $\nu = 0.2$ and the ultimate strain of the CF sheet $\varepsilon_{Lk}$ (Section 8.1.3):

$$\varepsilon_{juk} = \alpha_r \cdot \alpha_T \cdot \alpha_F \cdot \alpha_E \cdot \alpha_Z \cdot \varepsilon_{Lk} - \alpha_k \cdot \nu \cdot \varepsilon_{cc}(\Delta t)$$

$$\varepsilon_{juk} = 0.25 \cdot 0.7 \cdot 1.0 \cdot 1.0 \cdot 0.75 \cdot 15.2 - 1.5 \cdot 0.2 \cdot 0.44 = 1.863 \text{ mm/m}$$

### 8.5.3 Distribution of transverse compression

The following factors take into account the non-uniform distribution of the transverse compression over the cross-section owing to the different areas of influence of the confining reinforcement made up of reinforcing steel and CF sheet. In addition, we require the effective thickness of the inner link reinforcement $t_{w,eff}$ to DAfStb guideline [1, 2] part 1 Eq. (RV 6.92) and the diameter of the core cross-section confined by the reinforcing steel $D_c$ to DAfStb guideline part 1 Fig. RV 6.13:

$$t_{w,eff} = \frac{a_{sw}}{2 \cdot s_w} = \frac{5.23 \text{ cm}^2/\text{m}}{2} = 0.262 \text{ mm}$$

$$D_c = D - 2 \cdot c - 2 \cdot t_{w,eff} = 500 - 2 \cdot 30 - 2 \cdot 0.262 = 439.5 \text{ mm}$$

It is thus possible to calculate the factors for the non-uniform distribution of the transverse compression according to DAfStb guideline [1, 2] part 1 Eqs. (RV 6.90) and (RV 6.91):

$$p_1 = E_{jl} \cdot \varepsilon_{juk} = 690 \cdot 0.001863 = 1.29$$

$$p_2 = \frac{2 \cdot \left(E_L \cdot t_L \cdot \varepsilon_{juk} + t_{w,eff} \cdot f_{wyk}\right) - p_1 \cdot c}{D_c + c}$$

$$p_2 = \frac{2 \cdot (230\,000 \cdot 0.75 \cdot 0.001863 + 0.262 \cdot 500) - 1.29 \cdot 30}{439.5 + 30} = 1.84$$

The reduction in transverse compression as a result of the different areas of influence of the confining reinforcement to DAfStb guideline [1, 2] part 1 Eq. (RV 6.92) is required for the next calculations:

$$\Delta p = p_1 - \frac{2 \cdot E_L \cdot t_L \cdot \varepsilon_{juk} - (p_1 + p_2) \cdot c}{D_c}$$

$$= 1.29 - \frac{2 \cdot 230\,000 \cdot 0.75 \cdot 0.001863 - (1.29 + 1.84) \cdot 30}{439.5} = 0.041$$

### 8.5.4 Multi-axial stress state in concrete

The characteristic compressive strength of the confined concrete is calculated using DAfStb guideline [1, 2] part 1 Eq. (RV 6.65):

$$f_{cck} = f_{ck} + [k_1] \cdot \left[ E_{jl} \cdot \varepsilon_{juk} + \left(\rho_{wy} \cdot f_{wyk} - \Delta p\right) \cdot \left( \frac{D_c - \frac{s_w}{2}}{D} \right)^2 \right]$$

$$f_{cck} = 30 + 2 \cdot \left[ 690 \cdot 0.001863 + (0.0012 \cdot 500 - 0.041) \cdot \left( \frac{439.5 - \frac{300}{2}}{500} \right)^2 \right]$$

$$= 32.95\ \text{N/mm}^2$$

The transverse reinforcement ratio to DAfStb guideline [1, 2] part 1 Eq. (RV 6.65) is used here.

$$\rho_{wy} = \frac{2 \cdot t_{w,eff}}{D_c} = \frac{2 \cdot 0.262}{439.5} = 0.0012$$

To simplify the calculations for the stress block used in the design, the parameter $f_{ck}^*$ for the simplified stress–strain curve is used according to DAfStb guideline [1, 2] part 1 Eq. (RV 6.93):

$$f_{ck}^* = f_{ck} + [k_1] \cdot \left[\rho_{wy} \cdot f_{wyk} - \Delta p\right] \cdot \left( \frac{D_c - \frac{s_w}{2}}{D} \right)^2$$

$$= 30 + 2 \cdot [0.0012 \cdot 500 - 0.041] \cdot \left( \frac{439.5 - \frac{300}{2}}{500} \right)^2 = 30.37\ \text{N/mm}^2$$

The stress block geometry factor $\alpha_1$ according to DAfStb guideline [1, 2] part 1 Eq. (RV 6.94) is also required:

$$\alpha_1 = 1.17 - 0.2 \cdot \frac{f_{cck}}{f_{ck}^*} = 1.17 - 0.2 \cdot \frac{32.95}{30.37} = 0.953$$

### 8.5.5 Calculation of column load-carrying capacity

The calculation to establish the load-carrying capacity of the column is carried out according to DAfStb guideline [1, 2] part 1 section 6.1.4.2 (RV 5) and Eq. (RV 6.63). To do this, the relative angle $\theta$ must be determined iteratively with the following equations. The relative angle $\theta$ is estimated as $\theta = 0.809$ for this example. Taking this angle, the axial load capacity of the column according to DAfStb guideline [1, 2] part 1 eq. (RV 6.62) is

$$N_{\mathrm{Rd}} = \frac{1}{\gamma_{\mathrm{LG}}} \cdot \theta \cdot \alpha_1 \cdot f_{\mathrm{cck}} \cdot A_{\mathrm{c}} \cdot \left(1 - \frac{\sin(2 \cdot \pi \cdot \theta)}{2 \cdot \pi \cdot \theta}\right) + \frac{1}{\gamma_{\mathrm{s}}} \cdot (\theta_{\mathrm{c}} - \theta_{\mathrm{t}}) \cdot f_{\mathrm{syk}} \cdot A_{\mathrm{s}}$$

Here, the relative angle $\theta_{\mathrm{c}}$, which describes the stress distribution in the distributed longitudinal reinforcing steel in compression, and the relative angle $\theta_{\mathrm{t}}$, which takes into account the stress distribution in the distributed longitudinal reinforcing steel in tension, are determined according to DAfStb guideline [1, 2] part 1 Eqs. (RV 6.94) and (RV 6.95) depending on $\theta$:

$$0 \leq \theta_{\mathrm{c}} = 1.25 \cdot \theta - 0.125 \leq 1$$

$$\theta_{\mathrm{c}} = 1.25 \cdot 0.809 - 0.125 = 0.886$$

$$0 \leq \theta_{\mathrm{t}} = 1.125 - 1.5 \cdot \theta \leq 1$$

$$\theta_{\mathrm{t}} = 1.125 - 1.5 \cdot \theta = -0.1 \Rightarrow \theta_{\mathrm{t}} = 0$$

With these values available it is now possible to calculate the axial load capacity of the column:

$$N_{\mathrm{Rd}} = \frac{1}{1.35} \cdot 0.809 \cdot 0.953 \cdot 32.95 \cdot 1964 \cdot 10^2 \cdot \left(1 - \frac{\sin(2 \cdot \pi \cdot 0.809)}{2 \cdot \pi \cdot 0.809}\right)$$
$$+ \frac{1}{1.15} \cdot (0.886 - 0) \cdot 500 \cdot 5890 = 6642.4\ \mathrm{kN}$$

It is also necessary to check whether the acting moment corresponds to the resistance to moment actions. The maximum acting moment according to second-order theory taking into account creep deformations is calculated from the first part of Eq. (RV 6.63) according to the DAfStb guideline [1, 2]:

$$M_{\mathrm{Ed}} = N_{\mathrm{Rd}} \cdot \left(e_{\mathrm{tot}} + \frac{l^2}{\pi^2} \cdot \xi_1 \cdot \xi_2 \cdot \phi_{\mathrm{bal}} \cdot K_{\varphi}\right)$$

To calculate the maximum acting moment, further variables are required, which are calculated below. The factor taking into account the decrease in the curvature of the member as the longitudinal compressive force rises is calculated according to DAfStb guideline [1, 2] part 1 Eq. (RV 6.97):

$$\xi_1 = \frac{N_{\mathrm{bal}}}{N_{\mathrm{Rk}}} = \frac{0.8 \cdot f_{\mathrm{cck}} \cdot A_{\mathrm{c}}}{N_{\mathrm{Rd}} \cdot \gamma_{\mathrm{LG}}} \leq 1$$

$$\xi_1 = \frac{0.8 \cdot 32.95 \cdot 1964 \cdot 10^2}{6642.4 \cdot 10^3 \cdot 1.35} = 0.58$$

The factor to allow for the geometry of the compression member and the strain in the confining reinforcement is determined according to DAfStb guideline [1, 2] part 1 Eq. (RV 6.98):

$$\xi_2 = 1.15 + 0.06 \cdot \rho_\varepsilon - (0.01 + 0.012 \cdot \rho_\varepsilon) \cdot \frac{l_0}{\mathrm{D}} \leq 1$$

$$\xi_2 = 1.15 + 0.06 \cdot 0.932 - (0.01 + 0.012 \cdot 0.932) \cdot \frac{3000}{500} = 1.08 \Rightarrow 1.0$$

The strain coefficient $\rho_\varepsilon$ according to DAfStb guideline [1, 2] part 1 Eq. (RV 6.99) is used here.

$$\rho_\varepsilon = \frac{\varepsilon_{\mathrm{juk}}}{\varepsilon_{\mathrm{c2}}} = \frac{1.863}{2.0} = 0.932$$

The maximum curvature of the confined cross-section is determined according to DAfStb guideline [1, 2] part 1 Eq. (RV 6.100):

$$\phi_{\mathrm{bal}} = 2 \cdot \frac{\varepsilon_{\mathrm{cu}} + \varepsilon_{\mathrm{yk}}}{D + D_{\mathrm{c}} - (2 \cdot \phi_{\mathrm{w}} + \phi_{\mathrm{s}})}$$

$$\phi_{\mathrm{bal}} = 2 \cdot \frac{0.00479 + 0.0025}{500 + 439.5 - (2 \cdot 10 + 25)} = 1.63 \cdot 10^{-5}$$

In the above equation the ultimate strain of the confined concrcte $\varepsilon_{\mathrm{cu}}$ and the strain in the longitudinal reinforcing steel upon reaching the characteristic yield strength $\varepsilon_{\mathrm{yk}}$ are required according to DAfStb guideline [1, 2] part 1 Eqs. (RV 6.101) and (RV 6.102):

$$\varepsilon_{\mathrm{cu}} = \varepsilon_{\mathrm{c2}} \cdot \left(1.75 + 19 \cdot \frac{E_{\mathrm{jl}} \cdot \varepsilon_{\mathrm{juk}}}{f_{\mathrm{cm}}}\right) = 2.0 \cdot \left(1.75 + 19 \cdot \frac{690 \cdot 1.863 \cdot 10^{-3}}{38}\right)$$

$$= 4.79\,\mathrm{mm/m}$$

$$\varepsilon_{\mathrm{yk}} = \frac{f_{\mathrm{syk}}}{E_{\mathrm{s}}} = \frac{500}{200\,000} = 2.5\,\mathrm{mm/m}$$

The factor $K_\varphi$ takes into account the increase in the curvature due to the creep processes over time and is calculated according to DIN EN 1992-1-1 [20] Eq. (RV 5.37):

$$K_\varphi = 1 + \beta \cdot \varphi_{\mathrm{ef}} \geq 1$$

$$K_\varphi = 1 + 0.34 \cdot 0.45 = 1.15$$

The coefficient $\beta$ to DIN EN 1992-1-1 [20] section 5.8.8.3 (4) and the effective creep coefficient $\varphi_{\mathrm{ef}}$ to DAfStb guideline [1, 2] part 1 Eq. (RV 6.103) are used here.

$$\beta = 0.35 + \frac{f_{ck}}{200} - \frac{\lambda}{150}$$

$$\beta = 0.35 + \frac{30}{200} - \frac{23.9}{150} = 0.34$$

$$\varphi_{ef} = [k_7] \cdot \frac{16.8}{\sqrt{f_{cm}\left[\mathrm{N/mm^2}\right]}} \cdot \beta_{0,k} \frac{M_{0,Eqp}}{M_{Ed}}$$

$$\varphi_{ef} = 0.39 \cdot \frac{16.8}{\sqrt{38}} \cdot 1.0 \frac{20.7}{48.5} = 0.45$$

Using these variables it is possible to calculate the maximum acting moment:

$$M_{Ed} = N_{Rd} \cdot \left( e_{tot} + \frac{l^2}{\pi^2} \cdot \xi_1 \cdot \xi_2 \cdot \phi_{bal} \cdot K_\phi \right)$$

$$= 6642.4 \cdot 10^3 \left( 7.5 + \frac{3000^2}{\pi^2} \cdot 0.58 \cdot 1 \cdot 1.63 \cdot 10^{-5} \cdot 1.15 \right) \cdot 10^{-6} = 115.7\,\mathrm{kNm}$$

It is now necessary to calculate the resistance of the cross-section with the second part of Eq. (RV 6.63) according to the DAfStb guideline [1, 2]:

$$M_{Rd} = \frac{1}{\gamma_{LG}} \cdot \frac{2}{3} \cdot \alpha_1 \cdot f_{cck} \cdot A_c \cdot \frac{D}{2} \cdot \left( \frac{\sin^3(\pi \cdot \theta)}{\pi} \right)$$
$$+ \frac{1}{\gamma_s} \cdot f_{syk} \cdot A_s \cdot \frac{D}{2} \cdot \frac{\sin(\pi \cdot \theta_c) + \sin(\pi \cdot \theta_t)}{\pi}$$

$$M_{Rd} = \frac{1}{1.35} \cdot \frac{2}{3} \cdot 0.953 \cdot 32.95 \cdot 1964 \cdot 10^2 \cdot \frac{500}{2} \cdot \left( \frac{\sin^3(\pi \cdot 0.809)}{\pi} \right) \cdot 10^{-6}$$
$$+ \frac{1}{1.15} \cdot 500 \cdot 5890 \cdot \frac{500}{2} \cdot \frac{\sin(\pi \cdot 0.886) + \sin(\pi \cdot 0)}{\pi} \cdot 10^{-6} = 115.1\,\mathrm{kNm}$$

As the resistance is equal to the action, the relative angle $\theta$ chosen was correct.

$$M_{Rd} = 115.7\,\mathrm{kNm} \approx M_{Ed} = 115.1\,\mathrm{kNm}$$

As the acting axial load is less than the axial load resistance of the column, the column load-carrying capacity is satisfactory.

$$N_{Rd} = 6642.4\,\mathrm{kN} \geq N_{Ed} = 6469.8\,\mathrm{kN}$$

## 8.6 Serviceability limit state

In order to avoid unacceptable damage to the concrete microstructure at the serviceability limit state, DAfStb guideline [1, 2] part 1 section 7.2 (RV 15) specifies that the theoretical thickness of confining reinforcement necessary $t_L$ must comply with the following condition according to DAfStb guideline [1, 2] part 1 Eq. (RV 7.5):

$$t_L \leq \frac{D}{2 \cdot E_L \cdot \varepsilon_{juk}} \cdot \frac{1}{[k_1]} \cdot \left[\gamma_{LG} \cdot \left[\gamma_F \cdot ([k_8] - [k_9] \cdot f_{ck}) \cdot \left(\alpha_{cc} \cdot f_{ck} + \frac{A_s}{A_c} \cdot |\varepsilon_{c2}| \cdot E_s\right) - \frac{f_{syk}}{\gamma_s} \cdot \frac{A_s}{A_c}\right] - f_{ck}\right]$$

$$t_L \leq \frac{500}{2 \cdot 230\,000 \cdot 1.863 \cdot 10^{-3}} \cdot \frac{1}{2.0} \cdot \left[1.35 \cdot \left[1.43 \cdot (0.89 - 0.0044 \cdot 30) \cdot \left(0.85 \cdot 30 + \frac{58.9}{1964} \cdot 2.0 \cdot 200\right) - \frac{500}{1.15} \cdot \frac{58.9}{1964}\right] - 30\right]$$

$$t_L \leq 2.12 \text{ mm}$$

In this equation the permissible compressive strain in the concrete $\varepsilon_{c2}$ was determined according to DIN EN 1992-1-1 [20] Table 3.1 and the reduction factor for uniaxial concrete compressive strength $\alpha_{cc} = 0.85$ to DIN EN 1992-1-1/NA [21] section 31.6 (1). In addition, the weighted partial safety factor $\gamma_F$ for actions was calculated according to the contributions of the permanent and variable actions for the governing combination of forces and moments at the ultimate limit state:

$$\gamma_F = \frac{N_{Ed}}{G_k + Q_k} = \frac{6469.8}{2014.7 + 2500} = 1.43$$

As the thickness of confining reinforcement used is less than the maximum permissible thickness according to DAfStb guideline [1, 2] part 1 Eq. (RV 7.5), the design for serviceability is satisfactory.

$$t_L = 0.75 \text{ mm} \leq t_{L,max} = 2.12 \text{ mm}$$

# 9 Summary and outlook

This book has explained the design concept of the DAfStb guideline on the strengthening of concrete members with adhesively bonded reinforcement and illustrated this by way of examples. German [1] and English [2] editions of the DAfStb guideline as well as DAfStb publication No. 595 (commentary and examples) [58, 59] can be obtained from Beuth Verlag.

The DAfStb guideline is the first one in Europe to regulate the strengthening of concrete members with adhesively bonded reinforcement in the form of a supplement to the Eurocode. As it is planned to produce a document for this type of strengthening in a future Eurocode 2, the current DAfStb guideline can serve as a good starting point. Besides achieving a standardized method of design throughout Europe, another aim is to transfer approvals to the European level (European Technical Approvals, ETA).

To conclude, the authors would like to thank all the members of the subcommittee of the German Committee for Structural Concrete (DAfStb) for the cooperation in drawing up the guideline. Thanks also go to the sponsors of the research projects within the scope of drafting the guideline for their financial assistance.

*Strengthening of Concrete Structures with Adhesively Bonded Reinforcement: Design and Dimensioning of CFRP Laminates and Steel Plates*. First Edition. Konrad Zilch, Roland Niedermeier, and Wolfgang Finckh.

# References

1. DAfStb-RiLi VBgB: (2012) Verstärken von Betonbauteilen mit geklebter Bewehrung [strengthening of concrete members with adhesively bonded reinforcement]. Deutscher Ausschuss für Stahlbeton, 2012. Beuth, Berlin.

2. DAfStb-Guideline: Strengthening of concrete members with adhesively bonded reinforcement. Deutscher Ausschuss für Stahlbeton, 2012, English version, Beuth, Berlin, 2014.

3. Schäfer, H. (1996) Verstärken von Betonbauteilen – Sachstandsbericht [strengthening of concrete members – state-of-the-art report]. Schriftenreihe des DAfStb No. 467, Beuth, Berlin.

4. Jesse, F. and Kaplan, F. (2011) Verfahren für Biegeverstärkungen an Stahlbetonbauteilen [methods for flexural strengthening of RC members]. *Bautechnik*, **88**, 433–442.

5. Bergmeister, K. (2008) Ertüchtigung im Bestand – Verstärkung mit Kohlenstofffasern [upgrading existing buildings – strengthening with carbon fibres], in Beton-Kalender 2009 (eds K. Bergmeister, F. Fingerloos and J.-D. Wörner), Ernst & Sohn, Berlin, pp. 187–230.

6. Rostásy, F., Holzenkämpfer, P. and Hankers, C. (1995) Geklebte Bewehrung für die Verstärkung von Betonbauteilen [externally bonded reinforcement for strengthening concrete members], in Beton-Kalender 1996 Part 2, Ernst & Sohn, Berlin, pp. 547–576.

7. Zilch, K., Niedermeier, R. and Finckh, W. (2011) Sachstandbericht Verstärken von Betonbauteilen mit geklebter Bewehrung [adhesively bonded reinforcement – state-of-the-art report]. Schriftenreihe des DAfStb No. 591, Beuth, Berlin.

8. Zilch, K. and Finckh, W. (2013) Mechanischer Hintergrund der Wirkungsweise der aufgeklebten Bewehrung [mechanics background to principle of externally bonded reinforcement], in *Baustoff und Konstruktion*, Commemorative publication for Prof. Budelmann's 60th birthday (eds R. Nothnagel and H. Twelmeier), Springer, Berlin.

9. Zilch, K., Niedermeier, R. and Finckh, W. (2012) Praxisgerechte Bemessungsansätze für das wirtschaftliche Verstärken von Betonbauteilen mit geklebter Bewehrung – Verbundtragfähigkeit unter statischer Belastung [practical design methods for economic strengthening of concrete members with externally bonded reinforcement – bond strength under static loading]. Schriftenreihe des DAfStb No. 592, Beuth, Berlin.

10. Budelmann, H. and Leusmann, T. (2012) Praxisgerechte Bemessungsansätze für das wirtschaftliche Verstärken von Betonbauteilen mit geklebter Bewehrung – Verbundtragfähigkeit unter dynamischer Belastung [practical design methods for economic strengthening of concrete members with externally bonded

*Strengthening of Concrete Structures with Adhesively Bonded Reinforcement: Design and Dimensioning of CFRP Laminates and Steel Plates*. First Edition. Konrad Zilch, Roland Niedermeier, and Wolfgang Finckh.

reinforcement – bond strength under dynamic loading]. Schriftenreihe des DAfStb No. 593, Beuth, Berlin.

11. Zilch, K., Niedermeier, R. and Finckh, W. (2012) Praxisgerechte Bemessungsansätze für das wirtschaftliche Verstärken von Betonbauteilen mit geklebter Bewehrung Querkrafttragfähigkeit [practical design methods for economic strengthening of concrete members with externally bonded reinforcement – shear strength]. Schriftenreihe des DAfStb No. 594, Beuth, Berlin.

12. DIN 820-1: (2009) Standardisation – Part 1: Principles. Deutsches Institut für Normung, Beuth, Berlin.

13. Finckh, W., Zilch, K. and Niedermeier, R. (2010) DAfStb-Richtlinie: "Verstärken von Betonbauteilen mit geklebter Bewehrung – Bemessung" [DAfStb guideline: strengthening of concrete members with adhesively bonded reinforcement – design]. In: Fischer, O. (ed.): Proceedings, 14. Münchner Massivbau Seminar.

14. Krams, J., Kleist, A. and Kuhnen, L. (2011) Verstärken von Betonbauteilen mit geklebter Bewehrung. Neue Regelungen durch die gleichnamige DAfStb-Richtlinie [Strengthening of concrete members with adhesively bonded reinforcement, new regulations in DAfStb guideline]. Erhaltung von Bauwerken [maintenance of structures], 2nd Colloquium, 26 Jan.

15. Zilch, K., Niedermeier, R. and Finckh, W. (2011) New German guideline for strengthening concrete structures with adhesive-bonded reinforcement. In: Motavalli, M., Havranek, B., Saqan, E. (eds): Proc. of SMAR 2011, 1st Middle East Conference on Smart Monitoring, Assessment & Rehabilitation of Civil Structures, 8–10 Feb 2011, Dubai, UAE; Dübendorf, EMPA.

16. Budelmann, H. and Leusmann, T. (Mar 2011) Die neue DAfStb-Richtlinie "Verstärken von Betonbauteilen mit geklebter Bewehrung" [new DAfStb guideline "strengthening of concrete members with adhesively bonded reinforcement"]. Tragwerkverstärkung mit CFK-Lamellen und Faserverbundwerkstoffen.

17. Zilch, K., Finckh, W., Niedermeier, R. and Wiens, U. (2011) DAfStb-Richtlinie: Verstärken von Betonbauteilen mit geklebter Bewehrung – Teil 1: Bemessung und Konstruktion. [DAfStb guideline: strengthening of concrete members with adhesively bonded reinforcement – part 1: design]. *Bauingenieur*, **86**, 197–206.

18. Finckh, W. and Zilch, K. (14 Nov 2011) Bemessungskonzept nach der neuen deutschen Richtlinie. Weiterbildung am Abend: Biegeverstärkung von Stahlbeton mit Klebebewehrung. Die neue deutsche Richtlinie [design concept according to new German guideline; evening training session: flexural strengthening of RC with adhesively bonded reinforcement; new German guideline]. Dübendorf, Switzerland.

19. Welter, R. (2012) Verstärken mit CFK-Lamellen, Grundlagen für eine fehlerfreie und wirtschaftliche Bemessung [strengthening with CFRP strips, principles for trouble-free, economic design]. *Bautechnik*, **89**, 48–57.

20. DIN EN 1992-1-1: (2011) Eurocode 2: Design of concrete structures – Part 1-1: General rules and rules for buildings; German version EN 1992-1-1:2004S AC:2010. Deutsches Institut für Normung.

21. DIN EN 1992-1-1/NA: (2011) National Annex – Nationally determined parameters – Eurocode 2: Design of concrete structures – Part 1-1: General rules and rules for buildings. Deutsches Institut für Normung.

22. DIN-EN 1504-1: (2005) Products and systems for the protection and repair of concrete structures – Definitions, requirements, quality control and evaluation of conformity – Part 1: Definitions; German version. Deutsches Institut für Normung.

23. DAfStb-RiLi SIB: (2001) Schutz und Instandsetzung von Betonbauteilen [protection and maintenance of concrete members]. Deutscher Ausschuss für Stahlbeton.

24. DIN EN 1990: (2010) Eurocode: Basis of structural design; German version EN 1990:2002S A1:2005S A1:2005/AC:2010. Deutsches Institut für Normung.

25. DIN EN 1990/NA: (2010) National Annex – Nationally determined parameters – Eurocode: Basis of structural design. Deutsches Institut für Normung.

26. fédération internationale du béton (pub.): (2001) Externally bonded FRP reinforcement for RC structures. bulettin 14, Lausanne.

27. Blaschko, M. (2001) Zum Tragverhalten von Betonbauteilen mit in Schlitze eingeklebten CFK-Lamellen [structural behaviour of concrete members with near-surface-mounted CFRP strips]. Dissertation, Technische Universität München, Department of Concrete Structures.

28. Z-36.12-70: (2008) Verstärkung von Stahlbetonteilen durch mit dem Baukleber "StoPox SK 41" Schubfest aufgeklebte Kohlefaserlamellen "Sto S&P CFK Lamellen" nach DIN 1045-1:2008-08 [strengthening of RC members with carbon fibre "Sto S&P CFRP strips" bonded using "StoPox SK 41" adhesive for a shear-resistant connection to DIN 1045-1]. Deutsches Institut für Bautechnik.

29. Z-36.12-73: Verstärken von Stahlbetonbauteilen durch in Schlitze verklebte Kohlefaserlamellen Carboplus nach DIN 1045-1:2008-08 [strengthening of RC members with near-surface-mounted Carboplus carbon fibre strips to DIN 1045-1]. Deutsches Institut für Bautechnik (2009)

30. Daus, S. (2007) Zuverlässigkeit des Klebeverbundes von nachträglich verstärkten Betonbauteilen [reliability of adhesive bond of strengthened concrete members]. Dissertation, Darmstadt TU, Institute of Concrete Structures.

31. DAfStb-RiLi Sfb: (2012) Stahlfaserbeton [steel fibre-reinforced concrete]. Deutscher Ausschuss für Stahlbeton.

32. Pfeiffer, U. (2009) Experimentelle und theoretische Untersuchungen zum Klebeverbund zwischen Mauerwerk und Faserverbundwerkstoffen [experimental and theoretical studies of adhesive bond between masonry and fibre-reinforced materials]. University of Kassel.

33. Romani, M. (2002) Biegezugverstärkung von Brettschichtholz mit CFK- und AFK-Lamellen [flexural strengthening of glulam with CFRP and AFRP strips]. *Bautechnik*, **79**, 216–224.

34. Rizkalla, S., Dawood, M. and Schnerch, D. (2008) Development of a carbon fiber-reinforced polymer system for strengthening steel structures. *Composites Part A: Applied Science and Manufacturing*, **39**, 388–397.

35. Jesse, F. and Curbach, M. (2009) Verstärken mit Textilbeton [strengthening with textile-reinforced concrete], in Beton-Kalender 2010 (eds K. Bergmeister, F. Fingerloos and J.-D. Wörner), Ernst & Sohn, Berlin, pp. 459–565.

36. Andrä, H.-P., König, G. and Maier, M. (2001) Einsatz vorgespannter Kohlefaser-Lamellen als Oberflächenspannglieder [use of prestressed carbon fibre strips as surface-mounted prestressing tendons]. *Beton- und Stahlbetonbau*, **96**, 737–747.

37. Vorwagner, A., Burtscher, S.L., Grass, G. and Kollegger, J. (2010) Verstärkung mit vorgespannten eingeschlitzten Lamellen [strengthening using prestressed near-surface-mounted strips]. *Beton- und Stahlbetonbau*, **105**, 9–18.

38. Motavalli, M., Czaderski, C. and Pfyl-Lang, K. (2011) Prestressed CFRP for strengthening of reinforced concrete structures: recent developments at EMPA. *Switzerland. Journal of Composites for Construction*, **15**, 194–205.

39. Hülder, G., Dallner, C. and Ehrenstein, G. (2006) Aushärtung von Epoxidharzklebstoffen zur nachträglichen Verstärkung von Tragwerken mit CFK-Lamellen [curing of epoxy adhesives for strengthening structures with bonded CFRP strips]. *Bauingenieur*, **81**, 449–454.

40. Stark, B. (2003) Beispiel für den Nachweis der ausreichenden Tragfähigkeit von CFK-verstärkten Betonbauteilen im Brandfall [example for verifying adequate loadbearing capacity of CFRP-strengthened concrete members in fire]. *Bautechnik*, **80**, 393–399.

41. DIN EN 1992-1-2: (2010) Eurocode 2: Design of concrete structures – Part 1–2: General rules – Structural fire design; German version EN 1992-1-2:2004S AC:2008. Deutsches Institut für Normung.

42. DIN EN 1992-1-2/NA: (2010) National Annex – Nationally determined parameters – Eurocode 2: Design of concrete structures – Part 1–2: General rules – Structural fire design. Deutsches Institut für Normung.

43. Palmieri, A., Matthys, S. and Taerwe, L. (2011) Fire testing of RC beams strengthened with NSM reinforcement. In: Sen, R., Seracino, R., Shield, C., Gold, W. (eds.): ACI SP-275: Fiber-Reinforced Polymer Reinforcement for Concrete Structures, 10th International Symposium (FRPRCS 10).

44. DIN EN 13670: (2011) Execution of concrete structures; German version EN 13670:2009. Deutsches Institut für Normung. Beuth, Berlin.

45. DBV-Merkblatt: (2008) Bauen im Bestand – Beton und Betonstahl [construction works in existing buildings – concrete and reinforcing steel]. Deutscher Beton- und Bautechnik-Verein.

46. Schnell, J., Loch, M. and Zhang, N. (2010) Umrechnung der Druckfestigkeit von zwischen 1943 und 1972 hergestellten Betonen auf charakteristische Werte [conversion of compressive strengths of concretes produced between 1943 and 1972 to characteristic values]. *Bauingenieur*, **85**, 513–518.

47. BVBS NaReRiLi: (2011) Richtlinie zur Nachrechnung von Straßenbrücken im Bestand (Nachrechnungsrichtlinie) [guideline for re-analysing existing road bridges). Federal Ministry of Transport, Building & Urban Development.

48. Schnell, J., Loch, M., Zilch, K. and Dunkelberg, D. (2012) Erläuterungen und Hintergründe zu den Werkstoffkennwerten der Nachrechnungsrichtlinie für bestehende Straßenbrücken aus Beton [explanation of and background to material properties specified in structural assessment provisions for older concrete road bridges]. *Bauingenieur*, **87**, 15–23.

49. Rehm, G., Nürnberger, U. and Frey, R. (1981) Zur Korrosion und Spannungsrisskorrosion von Spannstählen bei Bauwerken mit nachträglichem Verbund [corrosion and stress corrosion cracking of prestressing steels in post-tensioned concrete]. *Bauingenieur*, **56**, 275–281.

50. Wölfel, E. (1992) Einzelne Spannbetonbauwerke möglicherweise durch verzögerte Spannstahlbrüche gefährdet [individual prestressed concrete structures possibly at risk of delayed prestressing steel failures]. *Beton- und Stahlbetonbau*, **87**, 155–156.

51. BVBS HAWe SpRK: (2011) Handlungsanweisung zur Überprüfung und Beurteilung von älteren Brückenbauwerken, die mit vergütetem, spannungsrisskorrosionsgefährdetem Spannstahl erstellt wurden [instructions for checking and assessing older concrete bridges built with quenched and tempered prestressing steel at risk of stress corrosion cracking]. Ministry of Transport, Building & Urban Development.

52. Lingemann, J. (2010) Zum Ankündigungsverhalten von älteren Brückenbauwerken bei Spannstahlausfällen infolge von Spannungsrisskorrosion [advance warning behaviour of older bridge structures with prestressing steel failures due to stress corrosion cracking]. Dissertation, Technische Universität München, Department of Concrete Structures.

53. Marzahn, G., Maurer, R. and Zilch, K. (2012) Die Nachrechnung von bestehenden Straßenbrücken aus Beton [re-analysis of existing concrete road bridges], in Beton-Kalender 2013 (eds K. Bergmeister, F. Fingerloos and J.-D. Wörner), Ernst & Sohn, Berlin.

54. Zilch, K., Niedermeier, R. and Finckh, W. (2012) Querkrafttragfähigkeit von historisch mit Betonstabstahl bewehrten und mit geklebter Bewehrung

biegeverstärkten Betonbauteilen [shear strength of legacy concrete members reinforced with steel bars and flexural strengthening in the form of externally bonded reinforcement]. Research report, Technische Universität München, department of Concrete Structures.

55. Zilch, K., Niedermeier, R. and Finckh, W. (2012) Verankerung von Stahllaschen im Bereich von Biegemomenten mit wechselnden Vorzeichen [anchorage of steel plates in the region of bending moments with changing signs]. Research report, Technische Universität München, department of Concrete Structures.
56. Niedermeier, R. (2011) Verstärkung von Stahlbetondruckgliedern durch Umschnürung [confinement methods for strengthening RC columns]. Professorial thesis, Technische Universität München, Department of Concrete Structures.
57. Finckh, W. (2012) Einfluss bauteilspezifischer Effekte auf die Bemessung von mit CFK-Lamellen verstärkten Stahlbetonbauteilen [influence of member-specific effects on the design of RC members strengthened with CFRP strips]. Dissertation, Technische Universität München, Department of Concrete Structures.
58. Deutscher Ausschuss für Stahlbeton: (2013) Erläuterungen und Beispiele zur DAfStb-Richtlinie "Verstärken von Betonbauteilen mit geklebter Bewehrung" [Commentary on the DAfStb Guideline "Strengthening of concrete members with adhesively bonded reinforcement" with Examples]. Schriftenreihe des DAfStb No. 595, Beuth, Berlin.
59. Deutscher Ausschuss für Stahlbeton: (2014) Commentary on the DAfStb Guideline "Strengthening of concrete members with adhesively bonded reinforcement" with Examples. DAfStb Report No. 595, Beuth, Berlin.
60. Deutscher Ausschuss für Stahlbeton: (1972) Bemessung von Beton- und Stahlbetonbauteilen nach DIN 1045, Ausgabe Januar 1972, Biegung mit Längskraft, Schub, Torsion, Nachweis der Knicksicherheit [design of concrete and RC members to DIN 1045, Jan 1972 ed., bending plus axial force, shear, torsion, analysis of safety against buckling]. Schriftenreihe des DAfStb No. 220, Ernst & Sohn, Berlin.
61. Zilch, K., Jähring, A. and Müller, A. (2002) Zur Berücksichtigung der Nettobetonquerschnittsfläche bei der Bemessung von Stahlbetonquerschnitten mit Druckbewehrung [considering the net concrete area in the design of RC cross-sections with compression reinforcement]. In: Deutscher Ausschuss für Stahlbeton (pub.): Erläuterungen zu DIN 1045-1 [commentary to DIN 1045-1]. Schriftenreihe des DAfStb No. 525, Beuth, Berlin, pp. 147–161.
62. Zilch, K. and Zehetmaier, G. (2010) Bemessung im konstruktiven Betonbau nach DIN 1045-1 (Fassung 2008) und EN 1992-1-1 (Eurocode 2) [structural concrete design to DIN 1045-1 (2008 ed.) and EN 1992-1-1 (Eurocode 2)], Springer, Berlin/ Heidelberg.
63. Niedermeier, R. Gemischte Bewehrung bei klebearmierten Bauteilen [mixed reinforcement in members with externally bonded reinforcement]. In: Zilch, K. (ed.): Munich Concrete Structures Seminar, 1997. Massivbau heute und morgen –

Anwendungen und Entwicklungen [concrete structures today and tomorrow – applications and developments], pp. XV.1–XV.13.

64. Niedermeier, R. Verbundtragfähigkeit aufgeklebter Bewehrung [bond strength of externally bonded reinforcement]. In: Zilch, K. (ed.): 3rd Munich Concrete Structures Seminar (1999)

65. Neubauer, U. (2000) Verbundtragverhalten geklebter Lamellen aus Kohlenstofffaserverbundwerkstoff zur Verstärkung von Betonbauteilen [bond behaviour of externally bonded carbon fibre-reinforced material for strengthening concrete members]. Dissertation, Braunschweig TU. Institute of Building Materials, Concrete Construction & Fire Protection.

66. Niedermeier, R. (2001) Zugkraftdeckung bei klebearmierten Bauteilen [tension reinforcement curtailment in members with externally bonded reinforcement]. Dissertation, Technische Universität München, Department of Concrete Structures.

67. Niedermeier, R. (2005) Überprüfung der Verbundtragfähigkeit bei klebearmierten Stahlbetonbauteilen [checking the bond strength of RC members with externally bonded reinforcement]. Niedermeier, R. (ed.): Massivbau in ganzer Breite [whole range of concrete structures]. Commemorative publication for Univ.-Prof. Dr.-Ing. Konrad Zilch's 60th birthday. Springer, Berlin, pp. 223–230.

68. Finckh, W. and Zilch, K. (2012) Strengthening and rehabilitation of reinforced concrete slabs with carbon-fiber reinforced polymers using a refined bond model. *Computer-Aided Civil and Infrastructure Engineering*, **27** (5), 333–346.

69. Finckh, W. and Zilch, K. (2012) Influence of member-specific effects the bond force transfer of reinforced concrete strengthened with externally bonded CFRP-strips, In: Bond in Concrete, Brescia.

70. Finckh, W. and Zilch, K. (2011) Influence of the curvature on the bond force transfer of EBR. In: Sen, R., Seracino, R., Shield, C., und Gold, W. (eds): ACI SP-275: Fiber-Reinforced Polymer Reinforcement for Concrete. Structures 10th International Symposium (FRPRCS 10), Tampa, USA.

71. Finckh, W., Zilch, K. and Ellinger, M. (2011) Evaluating the bond stress relationship of externally bonded CFRP-strips at the intermediate crack element. In: Motavalli, M., Havranek, B., und Saqan, E. (eds): Proceedings of SMAR 2011, the 1st Middle East Conference on Smart Monitoring, Assessment and Rehabilitation of Civil Structures. 8–10 February 2011, Dubai.

72. DIN EN 1542: (1999) Products and systems for the protection and repair of concrete structures – Test methods – Measurement of bond strength by pull-off; German version EN 1542:1999. Deutsches Institut für Normung. Beuth, Berlin.

73. DIN 1048-2: (1991) Testing concrete; testing of hardened concrete (specimens taken in situ). Deutsches Institut für Normung. Beuth, Berlin.

74. Noakowski, P. (1988) Nachweisverfahren für Verankerung, Verformung, Zwangsbeanspruchung und Rissbreite. Kontinuierliche Theorie der Mitwirkung

des Betons auf Zug. Rechenhilfen für die Praxis [method for verifying anchorage, deformation, restraint stresses and crack width; continuous theory of concrete's contribution in tension; practical design aids]. Schriftenreihe des DAfStb No. 394, Beuth, Berlin.

75. Zehetmaier, G. (2006) Zusammenwirken einbetonierter Bewehrung mit Klebearmierung bei verstärkten Betonbauteilen [interaction of embedded reinforcement and externally bonded reinforcement in strengthened concrete members]. Dissertation, Technische Universität München, Department of Concrete Structures.

76. Zilch, K., Niedermeier, R. and Finckh, W. (2010) Bauteilspezifische Effekte auf die Verbundkraftübertragung von mit aufgeklebten CFK-Lamellen verstärkten Betonbauteilen [specific structural effects on the bond force transfer of reinforced concrete strengthened with externally bonded CFRP strips]. *Bauingenieur*, **85**, 97–104.

77. Niedermeier, R. and Zilch, K. (2001) Zugkraftdeckung bei klebearmierten Bauteilen [tension reinforcement curtailment in members with externally bonded reinforcement]. *Beton- und Stahlbetonbau*, **96**, 759–770.

78. Zilch, K., Zehetmaier, G. and Niedermeier, R. (2004) Zugkraftdeckung im Bereich von Endauflagern bei klebearmierten Biegebauteilen [tension reinforcement curtailment in the region of end supports in members in bending with externally bonded reinforcement]. Research report, Technische Universität München, Department of Concrete Structures.

79. Zehetmaier, G. and Zilch, K. (2003) Interaction between internal bars and externally FRP Reinforcement in RC Members, Proceedings of the sixth international Symposium on FRP reinforcement for concrete structures (FRPRCS-6), Singapore, pp. 397–406.

80. Holzenkämpfer, P. (1994) Ingenieurmodelle des Verbunds geklebter Bewehrung für Betonbauteile [engineering model of bond of externally bonded reinforcement for concrete members]. Dissertation, Braunschweig TU. Institute of Building Materials, Concrete Construction & Fire Protection.

81. Holzenkämpfer, P. (1997) Ingenieurmodelle des Verbunds geklebter Bewehrung für Betonbauteile [engineering model of bond of externally bonded reinforcement for concrete members]. Schriftenreihe des DAfStb No. 473, Beuth, Berlin.

82. Husemann, U. (2009) Erhöhung der Verbundtragfähigkeit von nachträglich aufgeklebten Lamellen durch Bügelumschließungen [increasing the bond strength of retrofitted externally bonded strips by means of shear wrapping]. Dissertation, Braunschweig TU. Institute of Building Materials, Concrete Construction & Fire Protection.

83. Husemann, U. and Budelmann, H. (2009) Increase of the Bond Capacitiy of Externally Bonded CFRP-Plates on RC-Structures Due to Self-Induced Contact Pressure, In: Proceedings of the 9th International Symposium on Fibre Reinforced Polymers in Reinforced Concrete Structures (FRPRCS 9), Sydney, Australia.

84. Finckh, W. and Zilch, K. (2010) Shear capacity of flexural strengthened reinforced concrete structures with CFRP materials, In: Proceedings of the Fifth International Conference on FRP Composites in Civil Engineering, Vol. II FRP Strengthening Structures, 27. – 29. September 2010, Peking, China, pp. 794–797.

85. Finckh, W. and Zilch, K. (2012) Influence of shear crack offsets on the bond behavior of EBR at the intermediate crack element. In: The 6th International Conference on FRP Composites in Civil Engineering - CICE 2012, Rom.

86. Zilch, K. and Borchert, K. (2006) Geklebte Übergreifungsstöße von Schubbügeln und Stahllaschen [bonded laps for shear straps and steel plates]. Research report.

87. Jansze, W. (1997) Strengthening of reinforced concrete members in bending by externally bonded steel plates. Design for beam shear and plate anchorage. Dissertation, Delft University of Technology.

88. Leusmann, T. and Budelmann, H. (2012) Fatigue design concept for externally CFRP-plates, In: The 6th International Conference on FRP Composites in Civil Engineering - CICE 2012, Rom.

89. Budelmann, H., Husemann, U., Block, K. and Dreier, F. (2006) Einfluss von nicht vorwiegend ruhender Belastung auf die Verbundtragfähigkeit von CFK-Lamellen zur Bauteilverstärkung [influence of predominantly non-static loads on the bond strength of CFRP strips for strengthening members]. Research report.

90. Budelmann, H. and Husemann, U. (2008) Einfluss von nicht vorwiegend ruhender Belastung auf die Verbundtragfähigkeit von CFK-Lamellen zur Bauteilverstärkung [influence of predominantly non-static loads on the bond strength of CFRP strips for strengthening members]. Research report.

91. Zehetmaier, G. and Zilch, K. (2008) Rissbildung und Rissbreitenbeschränkung bei Verstärkung mit CFK-Lamellen [formation of cracks and crack control with near-surface-mounted CFRP strips]. *Bauingenieur*, **83**, 19–26.

92. Zilch, K. and Reitmayer, C. (2012) Zur Verformungsberechnung nach Eurocode 2 mit Hilfsmitteln [calculating deformations of concrete structures according to Eurocode 2 using design aids]. *Bauingenieur*, **87**, 253–266.

93. Z-36.1-1: (1979) Schubfeste Klebeverbindung zwischen Stahlplatten und Stahlbetonbauteilen [shear-resistant adhesive connection between steel plates and RC members]. Deutsches Institut für Bautechnik, Berlin.

94. DIN 1045: (1988) Structural use of concrete – Design and construction. Deutsches Institut für Normung. Beuth, Berlin.

95. DIN EN 13791: (2008) Assessment of in situ compressive strength in structures and precast concrete components; German version EN 13791:2007. Deutsches Institut für Normung. Beuth, Berlin.

96. Bertram, D. (2000) Betonstahl, Verbindungselemente, Spannstahl [reinforcing steel, connecting elements, prestressing steel], in Beton-Kalender 2001 (ed. J.F. Eibl), Ernst & Sohn, Berlin, pp. 145–204.

97. DIN 488-2: (1986) Reinforcing steels – Reinforcing steel bars. Deutsches Institut für Normung. Beuth, Berlin.

98. DBV-Merkblatt: (2002) Bautechnik – Betondeckung und Bewehrung [building technology – concrete cover and reinforcement]. Deutscher Beton- und Bautechnik-Verein.

99. Blaschko, M. (2003) Bond Behaviour of CFRP Strips Glued into Slits, Proceedings of the sixth international Symposium on FRP reinforcement for concrete structures (FRPRCS-6), Singapore, pp. 205–214.

100. Borchert, K. (2009) Verbundverhalten von Klebebewehrung unter Betriebsbedingungen [bond behaviour of externally bonded reinforcement under operating conditions]. Schriftenreihe des DAfStb No. 575, Beuth, Berlin.

101. Zilch, K. and Finckh, W. (2012) Erweiterung der Verbundmodelle auf eingeschlitzte CFK-Bewehrung [extending the bond model to near-surface-mounted CFRP reinforcement]. Commemorative publication for Prof. Eligehausen's 70th birthday.

102. Eligehausen, R., Popov, E. and Bertero, V. (1982) Local bond stress–slip relationship of deformed bars under generalized excitations. Research report, Berkeley.

103. Eligehausen, R. and Mayer, U. (2000) Untersuchungen zum Einfluss der bezogenen Rippenfläche von Bewehrungsstäben auf das Tragverhalten von Stahlbetonbauteilen im Gebrauchs- und Bruchzustand [studies of the influence of the projected rib area of reinforcing bars on the structural behaviour of RC members at the serviceability and ultimate limit states]. Schriftenreihe des DAfStb No. 503, Beuth, Berlin.

104. fédération internationale du béton (pub.): (2000) Bond of reinforcement in concrete. Lausanne.

105. Model Code 2010-1: (2010) Model Code 2010 – First complete draft, vol. 1, *fib* bulletin 55, fédération internationale du béton, Lausanne.

106. Rehm, G. and Franke, L. (1982) Kleben im Konstruktiven Betonbau [adhesive joints in structural concrete]. Schriftenreihe des DAfStb No. 331, Ernst & Sohn, Berlin.

107. Rehm, G., Franke, L. and Zeus, K. (1980) Kunstharzmörtel und Kunstharzbetone unter Kurzeit- und Dauerstandbelastung [synthetic resin mortars and concretes subjected to short- and long-term loads]. Schriftenreihe des DAfStb No. 309, Ernst & Sohn, Berlin.

108. Franke, L. and Deckelmann, G. (2003) Die Biegebemessung von Stahlbetonbauteilen mit nachträglich aufgeklebter Bewehrung unter Gesichtspunkten einer ausreichenden Dauerhaftigkeit [design for bending of RC members with retrofitted externally bonded reinforcement from the point of view of adequate durability]. Commemorative publication for Prof. Dr.-Ing. Peter Schießl's 60th birthday. Schriftenreihe Baustoffe No. 2, Technische Universität München, pp. 181–187.

109. Kleist, A. and Krams, J. (2008) Nachträgliche Rissbreitenbeschränkung mit CFK-Lamellen [CFRP strips for subsequent crack width limitation]. *Beton- und Stahlbetonbau*, **1**, 38–39.

110. Considère, A. (1902) Résistance à la compression du béton armé du béton fretté, *Beton und Eisen 1*, pp. 2ff.

111. Considère, A. (1903) Résistance à la compression du béton armé du béton fretté, *Beton und Eisen 2*, pp. 49ff, 101ff.

112. Müller, K.F. (1978) Tragfähigkeit und Verformung wendelbewehrter Stahlbetonstützen unter mittiger Belastung [loadbearing capacity and deformation of RC columns with helical reinforcement under moderate loading]. *Beton- und Stahlbetonbau*, **73**, 124–128.

113. Menne, B. (1977) Zur Traglast der ausmittig gedrückten Stahlbetonstütze mit Umschnürungsbewehrung [ultimate load of eccentrically loaded RC column with confining reinforcement]. Schriftenreihe des DAfStb No. 285, Ernst & Sohn, Berlin.

114. DIN 1045: (1972) Concrete and reinforced concrete structures – Design and construction. Deutsches Institut für Normung.

115. DIN 1045: (1975) Ergänzende Bestimmungen zur DIN 1045 Beton und Stahlbetonbau. Bemessung und Ausführung (Jan 1972 ed.) [supplementary provisions for DIN 1045]. Plain & Reinforced Concrete Working Group (Deutscher Ausschuss für Stahlbeton) of Construction Standards Committee, Deutscher Normenausschuss e.V. (DNA).

116. Fischer, A. (2010) Bestimmung modifizierter Teilsicherheitsbeiwerte zur semiprobabilistischen Bemessung von Stahlbetonkonstruktionen im Bestand [determination of modified partial safety factors for semi-probabilistic design of existing RC structures]. Dissertation, University of Kaiserslautern.

117. Fingerloos, F. and Schnell, J. (2008) Tragwerksplanung im Bestand [structural engineering for the building stock], in Beton-Kalender 2009 (eds K. Bergmeister, F. Fingerloos and J.-D. Wörner), Ernst & Sohn, Berlin, pp. 1–57.

118. DIN 18551: (2005) Shotcrete – Specification, production, design and conformity. Deutsches Institut für Normung.

119. ACI 440.2R-08: (2008) Guide for the design and construction of externally bonded FRP systems for strengthening concrete structures. American Concrete Institute.

120. CS-TR 55: (2004) Design guidance for strengthening concrete structures using fibre composite materials, 2nd ed., Concrete Society.

121. CNA/CSA S806-02: (2002) Design and construction of building components with fibre-reinforced polymers. Canadian Standards Association.

122. SIA 166: (2004) Klebebewehrung [reinforcement attached with adhesive]. Schweizerischer Ingenieur- und Architektenverein,.

123. CNR-DT 200: (2004) Guide for design and construction of externally bonded FRP systems for strengthening existing structures. Advisory Committee on Technical Recommendations for Construction, Rome.

124. Müller, K.F. (1975) Beitrag zur Berechnung der Tragfähigkeit wendelbewehrter Stahlbetonsäulen [contribution to calculating the loadbearing capacity of RC columns with helical reinforcement]. Dissertation, Technische Universität München, Department of Concrete Structures.

125. Sheikh, S.A. and Uzumeri, S.M. (1980) Strength and ductility of tied concrete columns. *Journal of Structural Engineering*, **105**, 1079–1102.

126. Parvin, A. and Schroeder, J.M. (2008) Investigation of eccentrically loaded CFRP-confined elliptical concrete columns. *Journal of Composites for Construction*, **12**, 93–101.

127. Yan, Z., Pantelides, C. and Reaveley, L.D. (2005) Shape modification with expansive cement concrete for confinement with FRP composites. 7th International Symposium on Fiber-Reinforced Polymer Reinforcement for Reinforced Concrete Structures (FRPRCS 7), ACI SP -230.

128. Tastani, S.P., Pantazopoulou, S.J., Zdoumba, D. *et al.* (2006) Limitations of FRP jacketing in confining old-type reinforced concrete members in axial compression. *Journal of Composites for Construction*, **10**, 13–25.

129. Nikitas, A. (2008) Zeitstandfestigkeit geklebter Übergreifungsstöße mit CFK-Gelegen [creep rupture strength of bonded laps with CF sheets]. Master's thesis, Technische Universität München, Department of Concrete Structures.

130. Lanig, N. (1988) Langzeitverhalten von Beton bei mehrachsiger Beanspruchung [long-term behaviour of concrete under multi-axial loading]. Dissertation, Technische Universität München, Department of Concrete Structures.

131. Xiao, Y. and Wu, H. (2000) Compressive behaviour of concrete confined by carbon fiber composite jackets. *Journal of Materials in Civil Engineering*, **12**, 139–146.

132. Eid, R. and Paultre, P. (2008) Analytical model for FRP-confined circular reinforced concrete columns. *Journal of Composites for Construction*, **12**, 541–552.

133. Lam, L. and Teng, J.G. (2003) Design-oriented stress–strain model for FRP-confined concrete. *Construction and Building Materials*, **17**, 471–489.

134. Leonhardt, F. and Mönnig, E. (1984) Vorlesungen über Massivbau. Teil 1: Grundlagen zur Bemessung im Stahlbeton [concrete structures lectures, part 1: RC design principles]. Springer, Berlin.

135. Jiang, T. (2008) FRP-confined RC columns: analysis, behavior and design. Dissertation, Hong Kong Polytechnic University.

136. Fitzwilliam, J. and Bisby, L.A. (2006) Slenderness effects on circular FRP-wrapped reinforced concrete columns: Proc. of 3rd International Conference on FRP Composites in Civil Engineering – CICE, pp. 499–502.

137. Ranger, M. and Bisby, L. (2007) Effects of load eccentricities on circular FRP-confined reinforced concrete structures. In: Triantafillou, T.C. (ed.): 8th

International Symposium on Fiber-Reinforced Polymer Reinforcement for Concrete Structures (FRPRCS 8).

138. Rüsch, H. (1960) Researches toward a general flexural theory for structural concrete. *ACI Journal*, **32**, 1–28.

139. Stöckl, S. (1981) Versuche zum Einfluss der Belastungshöhe auf das Kriechen des Betons [tests to establish how magnitude of loading influences creep of concrete]. Schriftenreihe des DAfStb No. 324, Ernst & Sohn, Berlin.

140. Al Chami, G., Thériault, M. and Neale, K.W. (2007) Time-dependent behaviour of CFRP-strengthened concrete columns and beams: CDCC 2007 – 3rd International Conference, pp. 235–242.

141. Naguib, W. and Mirmiran, A. (2002) Time-dependent behaviour of fiber-reinforced polymer-confined concrete columns under axial loads. *ACI Structural Journal*, **99**, 142–148.

142. CEB/FIP: (1993) Model Code 1990: Design code. Comité Européen du Béton, Fédération Internationale de la Précontrainte, Telford, London.

143. Wang, Y. and Zhang, D. (2009) Creep effect on mechanical behavior of concrete confined by FRP under axial compression. *Journal of Engineering Mechanics*, **135**, 1315–1322.

144. Berthet, J.-F., Ferrier, E., Hamelin, P. *et al.* (2006) Modelling of the creep behavior of FRP-confined short concrete columns under compressive loading. *Materials and Structures*, **39**, 53–62.

145. DIN 1045: (1959) Reinforced concrete structures. Deutsches Institut für Normung.

146. DIN 1045-1: (2001) Concrete, reinforced and prestressed concrete structures – Part 1: Design. Deutsches Institut für Normung. Beuth, Berlin.

# Index



*Strengthening of Concrete Structures with Adhesively Bonded Reinforcement: Design and Dimensioning of CFRP Laminates and Steel Plates*. First Edition. Konrad Zilch, Roland Niedermeier, and Wolfgang Finckh.